MIGUASHA PARK

FROM WATER TO LAND

Paleontology series
Directed by Marius Arsenault

Previous page: top view of a fossil skull of *Elpistostege watsoni*. Next page: fossil of *Bothriolepis canadensis*.

Photos : APM

Richard Cloutier
Isabelle Quinkal

MIGUASHA PARK
FROM WATER TO LAND

Translated by Peter Frost

MNH /   Parc de Miguasha

Illustration : F. Miville-Deschênes
Photo : APM

In my boyhood days, playing in front of our family home in Caplan, I could see far off to the west, about sixty kilometres away, the craggy outlines of the Notre Dame Mountains and Mont Saint-Joseph next to the town of Carleton. To my young eyes, this majestic backdrop was the end of the world! Often, with a pencil and a piece of paper in hand, I would try to sketch this landscape. It still remains one of the loveliest in the world—that of the south shore of the Gaspé Peninsula along Chaleur Bay. In my many sketches, I would pencil in a thin strip—a few millimetres thick—to depict a narrow line of land that straddled the horizon somewhere off beyond Carleton.

At the time, I had no idea that these few millimetres, lost in the blue of the sea, the mountains, and the sky, represented Point Miguasha and that this headland contained a fossil site that would play such a large role in my life. In time, I would learn that Miguasha was the dropoff point for a ferry service that brought American tourists to see Percé Rock or to go deer hunting and salmon fishing. Not until 1974 did I learn that Miguasha was a fossil fish site whose reputation had long spread far beyond the Gaspé Peninsula. I was 32 years old, was beginning a bachelor degree in geology, and was unaware that my corner of the country boasted a natural heritage site that outsiders had been excavating and studying for over a hundred years. It made me blush to think how little I knew about Miguasha.

That was all the motivation I needed. The next summer saw me walking along the beach in the company of a famous local collector, Euclide Plourde, who in his own way told me everything he knew about Miguasha. The same day, I bought two fossils for fifteen dollars…

That was my first encounter with the fossils of Miguasha.

Since then, over twenty-five years have passed and many other encounters have come and gone. One, with Richard Cloutier, was a valued moment for both of us. As we cast about for ways to share our common passion, i.e., the Miguasha site and paleontology, the idea of writing a book for the public at large, which would focus on this unique piece of geology, struck us as an absolute necessity! Other encounters were to follow, with André Martin and Isabelle Quinkal, of Publications MNH Inc., to make the project a reality. This book thus opens a new page and a new dimension on a fossil site that has become a Quebec provincial park and is now on UNESCO's World Heritage List.

Throughout its history, the Miguasha site has inspired many serious research projects and scientific publications. This book, Miguasha Park, from Water to Land, differs in that it is plainly written and seeks to make decades of scientific work available to one and all.

In closing, I hope this volume will enable others to look beyond the horizon and discover something as extraordinary as what I had the luck to learn about and help develop: a fossil site that has never ceased to teach us about the history of life and be a source of wonder for us all.

Marius Arsenault, paleontologist
Director, Miguasha Park,
UNESCO World Heritage Site

ACKNOWLEDGMENTS

To the teams of interpreter-guides who have succeeded each other at Miguasha Park over the years, I extend my thanks for their interest in and discussions of many aspects of paleontology at Miguasha. I wish to thank Marius Arsenault, Director of Miguasha Park (SÉPAQ), for the energy and passion he has devoted to developing the Miguasha site.

Several people read the manuscript in whole or in part with a critical eye. Their judicious comments have been much appreciated. They are: Marius Arsenault, Alain Blieck, Pierre Blier, Sophie Breton, Gisèle Cantin, Danielle Cyr, Thomas Grünbaum, Paul Lemieux, and Johanne Potvin.

To make the volume more attractive, many people provided photographs, drawings, and specimens. I sincerely thank them all: Eugene Balon, J. Russell Bodie, Robert L. Carroll, Robert Chabot, Jennifer A. Clack, Bill Clarke, Steve L. Cumbaa, Simon Conway-Morris, Rien Dam, Richard Day, Pierre Etcheverry, Richard Fournier, Pierre-Yves Gagnier, Patricia Gensel, Thomas Grünbaum, Hans J. Hoffman, Philippe Janvier, Simon Lamarre, Jean Leclerc, Ervins Luksevics, Andrew MacRae, John Maisey, Fred Marclay, Ben McHenry, Hans-Peter Schultze, William A. Shear, Bruno Vincent, Vincent Warwick, and Mark V.H. Wilson. For granting me access to the Miguasha Park photographic archives, I wish to thank the Miguasha Park management (SÉPAQ). I especially would like to thank François Miville-Deschênes for his artistic talents in bringing back to life these fishes from another age.

This project would not have been possible without the patience, professionalism, and dynamic assistance of the MNH team. Thanks to Isabelle Quinkal and André Martin, as well as to Peter Frost for the quality of the English translation.

Finally, I thank Olivier Cloutier. He lent an attentive ear and, through his childlike questions, opened up for me another vision of paleontology. And as only he can so admirably put it—when summing up evolution—life is forever !

Richard Cloutier

Preface 7

Acknowledgments 9

Introduction 13

Chapter I - Miguasha, a unique natural site 17
    UNESCO's world heritage 19
    History of the site : one and a half century of exploration .22
    Creation of Miguasha Park 30

Chapter II - Fossils and fossilization 33
    Fossils, a witness to the past 35
    Exceptional fossilization 43

Chapter III - Life and the Earth .47
    Evolution 48
    Evolution through the eyes of paleontology 59

    Our changing planet 76

    BOX : Did life arise on Earth
    or was the Earth "seeded"? .80

Chapter IV - The Devonian, " Age of Fishes " 81
    A crucial geological period in evolution 82
    What is a fish? 83

Chapter V - Miguasha, 370 million years ago **91**
     Quebec's rocks tell us... 92

     BOX : Miguasha, a Micmac place-name 96

     Miguasha today 97
     Miguasha in the Upper Devonian 100
     The fauna and flora of Miguasha 104

     BOX : *Latimeria* and the quest for living fossils 117

Chapter VI - What the Miguasha fossils tell us **125**
     How the fishes grew 126
     How the fishes fed 131
     How the fishes died 134

Conclusion
     Fossils and humans 137

Suggested readings **145**

Photo : APM

The quest for our origins begins amid stones.

The new millennium has already seen major new discoveries from the infinitely small to the infinitely large. After several years of work, the Human Genome Project—a collaborative effort involving fifteen research teams around the world—finally unveiled in June 2000 a complete map of the DNA sequences that are contained in each of our cells. A few days earlier, some astronomers announced that upon examining an interstellar dust cloud over 26,000 light years away they had found molecules from a special kind of sugar—one of the building blocks of DNA. Although these discoveries seem to involve two very different fields of science, they both shed light on how life first developed and, hence, on the basis of evolution itself !

At a time when great technological discoveries are being made with increasingly complex and sophisticated equipment, e.g., the Hubble telescope, some questions transcend this period of our history. Evolution, the origin of life, and the origin of Man are all themes that have long made us question and rethink our fundamental beliefs—striking deep as they do at the emotional and subjective bases of our existence. Although these questions have not always been formulated as clearly as they are now, humans very early on felt a need to know where they had come from, initially in terms of

Thousands of fossils. An impressive collection that is the basis for Richard Cloutier's work.

Photo : Canadian Geographic

generations and, later, in terms of millions of years. Scientists around the world are constantly trying to answer these questions by studying the life forms that have preceded us.

Paleontologists and evolutionary biologists are working hard to discover and decipher, bit by bit, the history of life on Earth. Every year, for over a century now, thousands of fossils have been collected and studied in order to understand our origins on this planet. A century, though, is just a tiny fraction of the history of life on Earth—which goes back several hundred million years.

The actinopterygian *Cheirolepis canadensis*.
(see page 115).

One of the most important sites in this search for our origins lies in Quebec, in the vicinity of Miguasha on the Gaspé Peninsula. It has opened a veritable window on life in the Devonian period. Though less impressive than dinosaur fossils, these remains are some 250 million years older! Each of us shares genes with some of the primitive fish that lived at Miguasha 370 million years ago and it is likely that the "father of our fathers," to borrow the title of Bernard Werber's latest novel, swam in Quebec's warm waters millions of years ago.

*A*rchaeopteris (see page 105).

MIGUASHA, A UNIQUE NATURAL SITE

The Miguasha cliffs border Chaleur Bay and contain thousands of fossils going back 370 million years.

Photo : APM

The Chaleur Bay area in Quebec contains a natural legacy several hundreds of millions years old. In 1842, fossils were discovered there— remains of animals and plants that attest to the evolution of life on earth. They have been studied since 1881 by scientists from around the entire world and particularly include fossils of archaic fish in sedimentary rocks from the Devonian geological period, commonly called the "Age of Fishes."

Location of Miguasha Park on the Gaspé Peninsula.

At that time, animal life was almost entirely aquatic—amphibians, reptiles, birds, and mammals having not yet appeared. Fish were the only vertebrates, i.e., back-boned animals. The Devonian saw these fish evolve and diversify, with several new groups appearing, dominating the aquatic environment, and giving rise to the first terrestrial vertebrates: tetrapods.

Because Miguasha documents in an exceptional manner this crucial stage in the history of life, it has become known worldwide in paleontology circles and has been added to the list of UNESCO's World Heritage Sites.

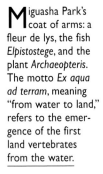

Miguasha Park's coat of arms: a fleur de lys, the fish *Elpistostege*, and the plant *Archaeopteris*. The motto *Ex aqua ad terram*, meaning "from water to land," refers to the emergence of the first land vertebrates from the water.

Photo : APM

UNESCO'S WORLD HERITAGE

The Convention concerning the Protection of the World Cultural and Natural Heritage of the United Nations Educational, Scientific and Cultural Organization—better known as the UNESCO World Heritage Convention—has a primary mission of protecting sites of importance to humanity, a heritage to be passed on to future generations.

Ratified in 1972, this convention covers both human-made cultural sites and natural sites. Among the cultural sites are the pyramids of Egypt, the Great Wall of China, the Taj Mahal in India, Mont Saint-Michel and its bay in France, the ancient fortress of Machu

Some cultural gems among UNESCO's World Heritage Sites. From top to bottom : Machu Picchu, the Sphinx and the pyramids of Egypt, and the Taj Mahal.

Picchu in Peru, and the historic district of Quebec City. They include monuments, structures, and sites that are of historical, aesthetic, archaeological, scientific, ethnological, or anthropological value.

Photos 1. 2. 3 : UNESCO

View from the Château Frontenac in Quebec City (below). Right: the Great Wall of China.

1

2

Some natural wonders among UNESCO's World Heritage Sites. Right: the Great Barrier Reef. The Galapagos Islands (below) whose rich fauna include giant tortoises

Natural heritage sites include noteworthy physical, biological, or geological formations, areas that are exceptional in terms of their scientific value, state of preservation, or natural beauty, and habitats of endangered animal and plant species. There were 138 of these sites as of December 2000. Among them are the Great Barrier Reef off Australia and the Galapagos Islands off Ecuador—a source of inspiration for Charles Darwin as he developed his theory of evolution.

3

Among the natural heritage sites are seven of paleontological interest. Miguasha was the sixth one to receive this global recognition. Two others are in Canada: Rocky Mountain Park and Dinosaur Provincial Park in Alberta. The first contains one of the most important sites for our understanding of the diversification of animal life on earth— the Burgess Shale Cambrian fossils in British Columbia.[1] The sec-

4

Photos 1, 2, 3, 4, 5 : UNESCO

5

ond is protected because of its wealth of fossil reptiles from the Cretaceous—35 dinosaur species have been discovered there.[2] In Europe, the Messel Pit site in Germany offers an exceptionally well-preserved find of 40 vertebrate species from the Eocene, an epoch of the Tertiary period.[3] The sites of Sterkfontein, Swartkrans, Kromdraai and its environs, in South Africa, have yielded specimens of *Australopithecus africanus* from 2 to 3 million years ago. The fossil sites of Riversleigh and Naracoorte in Australia document how marsupials and monotremes evolved during the Tertiary and Quaternary, their present-day descendants being kangaroos, koala bears, and duckbill platypuses. Finally, the nature parks of Ischigualasto-Talampaya in Argentina offer a complete sequence of Triassic sedimentary layers, including some of the oldest dinosaur remains known, going back 210 million years.

Photos 1, 2, 3, 4 : UNESCO

Views of Rocky Mountain Park (above) and Alberta's Dinosaur Provincial Park (left)

Left: a fossil of a small primitive horse from Germany's Messel Pit site (below)

1. see "The Explosion of Life" p. 64
2. see "The Age of Dinosaurs" p. 73
3. see "Geological Timescale" figure p. 58

A *Eusthenopteron foordi* specimen, chosen as a model for a stamp in the series "Prehistoric Life in Canada: The Age of Primitive Vertebrates" issued by Canada Post Corporation in 1991.

Photo : APM

Because of the relatively large number of Devonian fossil sites, a comparative study of over sixty of them was conducted prior to the request to add Miguasha Park to the list of world heritage sites. The study concluded that, of all the sites spread over the entire world, Miguasha had the most representative fauna and flora for this geological period. Its fossils stood out among all others; they are abundant, exceptionally well preserved, and representative of evolutionary lineages and past environments. On December 4, 1999, Miguasha Park joined UNESCO's list of World Heritage Sites.

HISTORY OF THE SITE: ONE AND A HALF CENTURIES OF EXPLORATION

DISCOVERY

The first studies of rocks on the north shore of the Restigouche River and its estuary go back to 1842, when

The Natural History Museum of Miguasha Park..

Photo : Armand Dubé

northern New Brunswick was being mapped. The British government commissioned a geological survey of the province from 1838 to 1842 under the auspices of Abraham Gesner, a physician by training who had returned to his first life passion— geology and fossil collecting. His survey took him as far as Miguasha, also called "Scaumenac Bay," where he discovered many fossils. In his 1843 report, he wrote:

Photo : APM

A second stamp from the same series, inspired this time by a fossil of *Archaeopteris halliana*.

"Crossing the harbour at Dalhousie, and on the Gaspe side of the Restigouche below Escuminac Bay, I found the shore lined with a coarse conglomerate.

Farther eastward the rocks are light blue sandstones and shales, containing remains of vegetables. (...) In these sandstones and shales, I found the remains of fishes, and a small species of tortoise with foot-marks."

The Nova Scotian Abraham Gesner first discovered the Miguasha fossils. He also invented kerosene and is considered to be a father of the oil industry.

Photo : APM

R. W. Ells

J. F. Whiteaves

Photos : APM

The right-hand illustration is an extract from *the Proceedings and Transactions of the Royal Society of Canada* for the year 1886. It illustrates the reconstitution by J.F. Whiteaves of the fish *Bothriolepis canadensis*.

Despite this mention, the Miguasha fossils lay undisturbed until their rediscovery in 1879 by a geologist from the Geological Survey of Canada, R.W. Ells. He came to collect samples from the Miguasha cliffs for a survey of the western part of Chaleur Bay and brought back around fifty fossil fish and several fossil plants. In 1880 and 1881, the Geological Survey of Canada investigated the fossil treasure trove through the pioneering excavations of R.W. Ells, A.H. Foord, and T.C. Weston. The fossil fish they collected were handed over to the Canadian paleontologist J.F.

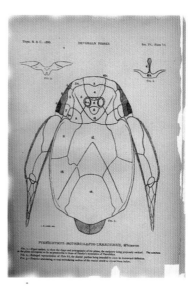

Whiteaves and the plant fossils to the paleobotanist J.W. Dawson, who in 1882 was the first to date the Miguasha specimens to the Upper Devonian. Whiteaves went on to describe six of the twenty fossil fish species now known from the site.

Photo : Bill Clarke, Musée régional de Dalhousie

ENGLISH AND AMERICAN SCIENTISTS ENTER THE FRAY

After Canada, other countries were quick to organize fossil expeditions to the Miguasha site. From Britain came the scientist Jex, who discovered several new species among the specimens he collected between 1887 and 1892. The specimens were then described by two famous British paleontologists, A. Smith Woodward and R.H. Traquair.

Meanwhile, there came in 1892 the American paleontologist E. D. Cope, known for his dinosaur work. He found that a Miguasha fossil fish, by the name of *Eusthenopteron foordi*, was morphologically related to the first four-legged vertebrates to live on land, the tetrapods.

The port of Dalhousie, New Brunswick, in the late 19th century

Photo : Redpath Museum

Knighted by Queen Victoria in 1884, Sir J.W. Dawson was one of the most important figures in North American geology. He published the first study on Miguasha fossil plants in 1880.

Photo : J.-P. Sylvestre

This cast of the Devonian fish *Eusthenopteron foordi* on a Miguasha beach takes us back many hundreds of millions of years.

From the late 19th century on, many American paleontologists followed in Cope's footsteps and visited the site to collect fossils. Their visits account for the existence of fossil material from Miguasha at many American museums, including the American Museum of Natural History in New York. Thus, from far and wide, investigators flocked to Miguasha in the fifty years following the first discoveries.

Euclide Plourde in 1964 (above) and Joseph Landry in 1937 (right): the same passion for fossils.

Photos : APM

Admittedly, they would all have found much less were it not for local collectors, such as the Plourde, Landry, and Roy families. These amateur fossil-hunters provided many American and European universities and museums with high-quality specimens.

THE SWEDISH SCHOOL

Foremost among the European pale-ontologists to have visited the Miguasha site is the Swedish professor Erik A. Stensiö. He came to work on-site in 1922, concentrating his energies on describing in detail the morphology of the armour-plated fish *Bothriolepis canadensis*. Back then, such a paleontological investigation could be done only on specimens from Miguasha. For the first time in vertebrate paleontology, a fossil site offered the pos-sibility of studying several hundred excep-tionally well-preserved specimens. These specimens provided scientists at the renowned Swedish school of vertebrate paleontology with exceptional material for study right up until the 1970s. Among them was the Swedish professor Erik Jarvik, who contributed greatly to enhancing the reputation of the Miguasha site through his lengthy work on *Eusthenopteron foordi*. Among other sci-entific papers, Jarvik wrote a 440-page description dealing mainly with the snout of this Devonian fish.

This prestigious Swedish school has produced many other big names in pale-oichthyology: Jean-Pierre Lehman, Hans C. Bjerring, Roger Miles, Tor Ørvig, Hans-Peter Schultze, Erol I. White, Philippe Janvier, and Jiri Zidek. Some are still actively doing research, and shedding light on Quebec's fossil fishes.

Photo : U. Samuelson, Natur Historiska Riksmuseet, Stockholm

Professor Erik A. Stensiö. Besides his work on *Bothriolepis*, he also described the jawless fish *Endeiolepis*. A mountain in Greenland, where remains of the first Devonian tetrapods were found, bears the name of Stensiö.

PALEONTOLOGY "*Made in Quebec*"

The year 1937 was a turning point in awakening Quebec's scientists to the Miguasha finds. While traveling through the Gaspé Peninsula, Father Léo-G. Morin and Reverend Joseph-Willie Laverdière, both geologists and, respectively, the first chairmen of the geology departments of the Université de Montréal and Université Laval in Quebec City, realized the importance of the fossil treasures at Miguasha. Numerous foreigners were already harvesting these finds, but Quebec's universities still had none. The two contacted the mining department of the Province of Quebec and it assigned René Bureau, a technician at the

Prof. Erik Jarvik (left) in 1991 giving Marius Arsenault, the director of Miguasha Park, the "Erik Stensiö" medal. Awarded by the Natural History Museum of Stockholm, it honours Arsenault's contribution to the development of Miguasha Park.

Right: father Léo-G. Morin in the late 1930s

Photos : APM

time, the task of harvesting the fossils that would form Quebec's first collection of Miguasha specimens. Bureau had the foresight to get the government authorities to put a stop to haphazard and potentially detrimental fossil excavations at the site.

In the 1960s, the Quebec ichthyologist Vianney Legendre, like René Bureau, sought to make the government authorities aware of the importance of the Miguasha fossil cliffs. Alongside his work on freshwater fish in Quebec, he had an abiding passion for the Miguasha fossil fish. For years he would be a fount of inspiration and knowledge for studies at Miguasha. Indeed, a lot of scientific research was going on! Over the last twenty years, on the initiative of Marius Arsenault, a team of Quebec paleontologists—Michel Belles-Isles, Pierre-Yves Gagnier, Daniel Vézina, and Richard Cloutier—have specialized in studying the paleontological and geological treasure trove of Miguasha. Through their discoveries, these graduates from institutions in Montreal, Paris, and Lawrence, Kansas have helped improve our understanding of these fossils that are unique in the world. Since 1980, Quebec paleontologists have written nearly 100 scientific papers on or referring to the Miguasha fossil fishes.

Abbé Joseph-Willie Laverdière

Photos : APM

CREATION OF MIGUASHA PARK

As early as the late 1930s, René Bureau had understood the importance of preventing uncontrolled collecting of the Miguasha fossils. His "Miguasha Project" bore fruit in the early 1970s, when the provincial government decided to acquire part of the fossil cliffs. Development of the site began in 1976 under the direction of the Université du Québec à Rimouski (UQAR). In 1978, the first interpretation centre, a rather modest building, was opened to the public and the site welcomed over 8,000 visitors in its first year.

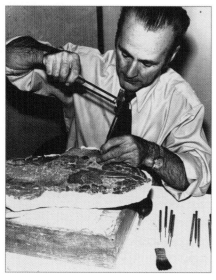

Beginning in 1937, René Bureau, who later became curator of the geology museum at Université Laval, strove unrelentingly to make known and protect the Miguasha fossils.

Photos : APM

The combined efforts of several players began to pay off. Bureau was one but there were also the citizens of Miguasha and Nouvelle, who submitted a brief to the Quebec government. In January 1985, after public hearings, the fossil site officially became one of the province's parks, thus ensuring that it would be protected and developed in keeping with the preservation of its natural heritage.

Marius Arsenault has run the park since its creation. He first came into contact with Miguasha in 1977 when, as a vertebrate paleontology student at the Université du Québec à Montréal, he was hired as a team leader by officials at the Université du Québec à Rimouski. Through his initiatives, many working relationships have been forged between the Miguasha Natural History Museum and many leading international paleontology organizations with a view to increasing what we know about the fossil fishes of the Devonian. In the summer of 1991, the park opened a new natural history museum and in the same year welcomed 63 paleontologists from fifteen different countries to the 7th International Symposium on the Studies of Early Vertebrates.

Finally, December 4, 1999 saw another milestone in the short modern-day history of the Miguasha fossil site. On that day, the World

Campus of the Université du Québec à Rimouski.

Photo : UQAR

VIIe SYMPOSIUM INTERNATIONAL
ÉTUDE DES VERTÉBRÉS INFÉRIEURS

Parc de Miguasha
9-22 juin 1991

Québec

Photo : APM

Heritage Committee recommended adding the natural site to UNESCO's World Heritage List. The commemorative plaque for the event, which today stands at the entrance to the Miguasha Natural History Museum, was officially unveiled on June 29, 2000. The inscription on it should encourage future genera-

The current Natural History Museum at Miguasha Park.

Centre: plaque commemorating the admission of Miguasha Park to the UNESCO list of World Heritage Sites. It was officially unveiled on June 29, 2000.

Below: *L'envol*, sculpture by Christopher Varady-Szabo in homage to the evolution of *Eusthenopteron*.

tions to perpetuate and share the fount of knowledge that the Miguasha fossils have provided us.

"Miguasha Park is considered the world's most representative paleontological site of the Devonian period called the "Age of Fishes." The site is of significance because it contains the greatest number and best-preserved fossil specimens of the lobe-finned fishes that gave rise to the tetrapods, the first terrestrial vertebrates."

Photos : APM

This specimen of *Eusthenopteron foordi* (foreground), preserved in three dimensions, measures around a metre in length. It has inspired many paleontologists in their hypotheses on the transition from water to land by vertebrates. In the background, enlarged many times, are fossils of the acanthodian *Triazeugacanthus affinis*. In some strata of the Escuminac Formation, over 600 specimens per square metre have been fossilized.

Photos : APM & The Cleveland Museum of Natural History

FOSSILS AND FOSSILIZATION

Our interest in fossils is apparently as old as our species. Prehistoric sites, some going back over 30,000 years, have repeatedly yielded assortments of corals and shells, sea urchins and ammonites—all fossils —gathered by the Neanderthals and the first *Homo sapiens*, the Cro-Magnons. This infatuation with fossils has continued unabated.

The Cyclops myth began with the discovery, 5,000 years ago, of bones preserved in a cave at the foot of Mount Etna.

The bones were thought to be human remains and the hole in the enormous skulls was believed to be the eye socket of one-eyed giants. In reality, it was the nasal orifice for the trunk of small elephants that lived in the early Quaternary on these Mediterranean islands.

In China 5,000 years ago, dinosaur fossils were thought to be dragon bones and were crushed for use in traditional medicine. Egyptians, Etruscans, and Greeks collected fossils and used them most often as jewellery, as did tribes of certain North American Indians who wore pendants made of fossils well before the Europeans arrived in North America. In Western Canada, sacred objects called *Iniskim* have been found in small medicine bags that belonged to the Blackfoot tribe. These *Iniskim* are, so it turns out, fossils of corals, ammonites, and other cephalopods that were painted red and wrapped in bison fur. They were supposed to bestow a special power on bison hunters who held them in their possession. There is a story that in 1857 an Indian guide refused to accompany the geologist and naturalist Henry Y. Hind—who wanted to check out a reported find of mammoth bones in Western Canada—for fear of approaching what he believed were the legendary bones of the Manitou, or Great Spirit.

For several centuries now, we have known that fossils offer an impressive record of past ages, insofar as we can decipher them. This work of deciphering that makes up the study of fossils, is one of the foundations of paleontology, the science that concerns itself with the evolution of life on earth.

FOSSILS, A WITNESS TO THE PAST

What is a fossil? How old are fossils? How does a fossil form? Many questions come to the surface, in the minds of the young and the not so young, when fossils are being discussed.

> **DEFINITION:**
> A fossil is said to be the remains or traces of an animal or plant that lived in the past.

What is a Fossil?

A fossil is an organism that, after its death, has been altered by complex chemical, physical, and biological phenomena, thus ensuring its preservation. A fossil is not just an imprint in a rock. It is a remnant or a trace of an organism that lived in the past.

Any plant or animal, in whole or in part, can fossilize, be it as small as bacteria measuring a few thousandths of a millimetre or as large as certain dinosaurs whose skeletons reach several dozen metres in length.

Two sides of the same coin: the remains and the imprint of *Bothriolepis canadensis* are both fossils in their own right.
A side view of the outline of this fish.

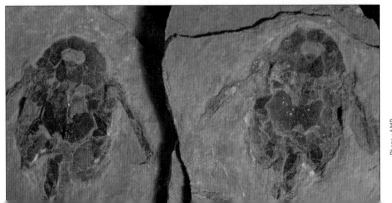

Photo : AMP
Illustration: R. Cloutier

Skeleton of the dinosaur *Tyrannosaurus rex*, around 3 metres tall, on display at the Royal Tyrrell Museum in Drumheller, Alberta.

Below: This flower-like ichnofossil from Miguasha is an example of *Gyrophyllites*. Each "petal" is a burrow dug by an invertebrate in the fine-grained sand in its search for food. The centre of the "flower" is the animal's main home, with the burrows radiating outwards.

A fossil can also be a trace left by an animal. We call this an ichnofossil. For example, the burrows made by aquatic organisms in sediments are just as much fossils as the animals themselves. Burrows, shelters, and tracks are all signs that attest to an animal's presence and that, in addition, can tell us about its behaviour. It is seldom possible, though, to associate a trace with the animal that left it.

In unique cases, fossilization can also preserve regurgitated food and animal excrement. These fossils are respectively referred to as regurgitates and coprolites.

Photos : R. Cloutier

Finally, a recent discovery has introduced a new line of fossil research. In 1996, a team of American and Canadian scientists found what they believe to be fos-

Fish coprolite preserved in the laminites of Miguasha.

Photo : R. Cloutier

sil bacteria inside a meteorite of Martian origin. The claim is still controversial. If ever confirmed, our definition of a fossil will have to include not only terrestrial remains but also extraterrestrial ones!

This electron microscope image (left) provides a close-up view of a meteorite of Martian origin. It has been enlarged several tens of thousands of times.

The arrows point to what scientists think are fossil bacteria, about 200 nm (nanometres) across, i.e., 200 billionths of a metre or ten times smaller than most bacteria on our planet.

Photo : McKay *et al.*, 1996, *Science*, vol 273, p 924.

HOW OLD ARE FOSSILS?

How long does an organism have to be dead before it is considered to be a fossil? Clearly, the fish found at Miguasha, going back 370 million years, are incon-

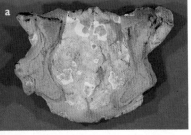

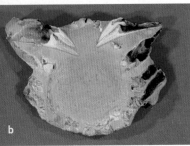

Photos : R. Cloutier
& R Chabot, UQAR

testably fossils. It is also accepted that the term "fossil" applies to the remains of 8,000 to 12,000 year-old fish, sea mammals, and invertebrates that have turned up in the clay, sand, and gravel of the St. Lawrence Lowlands, the Lower St. Lawrence, the Gaspé Peninsula, the Saguenay region, and the North Shore of the Gulf of St. Lawrence. Human skeletons buried in cemeteries, however, are not fossils, even though some may be hundreds of years old. Although it is difficult to set an exact limit beyond which bones and shells become fossils, fossilization may be said to have occurred once the complex chemical and biological transformations mentioned above have begun.

Another question concerns the age of fossils. How do we know that the Miguasha fossils are 370 million years old? Paleontology uses two approaches to fossil dating. One is "relative" and tries to match a fossil to a geological period, e.g., the Middle Frasnian. The other approach is "absolute" and tries to quantify the age of a rock, e.g., as 370 million years old. Relative dating, or biostratigraphy, relies on principles of correlation and similarity when comparing fossil materials, called "assemblages," from different sites. Two sites are considered to be synchronous, i.e., have the same age, if they have analogous assemblages. International charts of biostratigraphic sequences, based on specific groups of animal and plant fossils, serve as benchmarks for these relative dating methods.

Photos : R. Cloutier & R Chabot, UQAR

These fossils, around 10,000 years old, were found on the north shore of the Gulf of St. Lawrence, Quebec. Above: capelins, a fish about fifteen centimetres long that lives in cold waters.
Left-hand page: a snow crab seen from above (a) and below (b).

Thus, through comparison with European reference assemblages, it was possible to conclude that the Miguasha specimens date back to the Frasnian, a subdivision of the Devonian period.[1] But how can we narrow down the dating of Devonian rocks within this timeframe of 410 to 355 million years ago? This is where absolute dating methods come in, or geochronometry. These methods tell us how old the rock is, chiefly by using the radioactive properties of certain minerals. When a rock is newly formed,

1 see "Geological timescale" p. 58

The principle of absolute dating: the "parent" radioactive element initially present in the rock (here U^{238}) gradually decomposes into a "daughter" element (Pb^{206}). Its rate of decomposition, or disintegration, is not constant over time. It is rapid at first and then slows down later on, as may be seen from the diagram.

such as lava during a volcanic eruption, some of the chemical elements in the rock are radioactive and relatively unstable. The elements will be transformed over geological time at a regular rate into more stable elements. This is called decomposition. For instance, uranium (U^{238}) decomposes into lead (Pb^{206}). Such transformation follows predictable rules. The absolute age of a rock can then be determined by studying how fast its natural radioactivity decays in a laboratory, i.e., by calculating the respective amounts of its original unstable elements and its more stable, transformed elements.

Paleontologists, biostratigraphers, and geochemists have worked together to measure the absolute ages of different fossil assemblages all over the world,

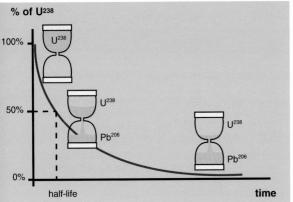

Half-life corresponds to the time when half of the parent element has decomposed. In the case of U^{238}, the half-life is 4.5 billion years.

thus providing a dating timescale that covers most geological periods. This method, which may seem long and complex, has found that the Devonian fossils at Miguasha are 370 million years old, give or take a few million.

HOW DOES A FOSSIL FORM?

Fossilization is mainly the result of a slow, continuous process. The overwhelming majority of fossils form in an aquatic environment, be it a lake, an estuary, or a sea, although steppes and deserts may also be conducive to fossilization. For fossilization to begin, the plant or animal must be quickly buried over several hours or days under several millimetres of sediment, e.g., sand, mud, or silt. This will prevent it from being destroyed by predators or scavengers, in the case of it being an animal.

If, in an aquatic environment, the animal was not rapidly buried after death, gases will form in its body through fermentation and it will float to the surface, thus decreasing its chances of fossilization. In the case of a fish, for example, its decomposing body will slowly break up and certain parts will become detached. Scales, skull bones, and portions of the backbone may end up being scattered over the sea or lake bottom. Whether complete or fragmentary, the specimens will become immersed in clay or sand sediments that will then harden to form shales or sandstones, two kinds of sedimentary rock. Thus begins the long process of fossilization.

The illustration below shows the successive stages in the process of fossilization of the lungfish *Fleurantia*: death of the animal, burial, fossilization.

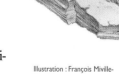

Illustration : François Miville-Deschênes

0.5 cm

Isolated scale from a Miguasha lungfish

Photo : R. Cloutier

This Belgian royal red marble, a metamorphic limestone, contains fossils of corals, brachiopods, and bivalves.

Photo : Stenen en marmers van Wallonie, C. Cnudde, J. J. Harotin. J. P. Majot.; A.A.M. Ed., 1990.

Most fossils are therefore found in sedimentary rock. Nonetheless, some may also appear in certain weakly metamorphosed rocks, whose structure has changed in response to high temperatures and pressures. Marble is a classic example of this kind of rock. Volcanic rocks are another, but only under exceptional conditions. Generally, lava and volcanic ash will destroy any trace of life because of their extremely high temperature. In the Western United States, however, remains of fauna and flora from the Tertiary period have been found preserved in volcanic ash. Another famous example: the discovery of human forms in the volcanic deposits that covered the Italian city of Pompeii when Mount Vesuvius erupted in 79 A.D.

In addition, fossilization is facilitated by an absence of free oxygen in the water, i.e., an anaerobic environment. Oxygen is essential to the survival of certain "gravedigger" bacteria, which play a role in the decomposition of dead organisms. An anaerobic environment inactivates these bacteria, thus helping preserve the plant or animal.

EXCEPTIONAL FOSSILIZATION

Of all the fossil sites that document the history of life, a few stand out because of the exceptional preservation of their fossils. In 1970, the German paleontologist Adolf Seilacher recognized a special class of fossil sites with exceptional preservation. He named them "Lagerstätten," *Lager* meaning layer and *Stätte* meaning site. This German name has since become part of the vocabulary of paleontology and is now widely used. The following examples clearly justify the Miguasha fossil cliffs being considered one of the rare Lagerstätten of Devonian vertebrates.

This beautifully preserved specimen of a porolepiform *Holoptychius jarviki* (left) attests to the exceptional quality of the preservation of Miguasha fossils.

Photo : American Museum of Natural History, New York.
Illustrations : R. Cloutier

Lagerstätte :
Lager : layer, Stätte : place
Term introduced in 1970 by the German paleontologist Adolf Seilacher.
Refers to a fossil site that exhibits exceptional preservation.

A unique scientific opportunity: (below) a complete specimen of *Escuminaspis laticeps*, about forty centimetres long, whose head and body are both composed of thousands of minuscule bony parts.

Photo : APM

Fossils are usually formed from the hard parts of organisms. The best-preserved parts are the mineralized shells and cases of invertebrates and the bones, teeth, and cartilage of vertebrates. Of the 14,000 fossil specimens from the Miguasha cliffs, most are the bony parts of fish. Several specimens, however, have been entirely preserved, i.e., all of the fish bones are present, sometimes none even being out of place. Not only have the shape and size of the bones remain unchanged during fossilization, but also, in the case of fossilization in three dimensions, the animal's position at the time of death has been preserved. This type of fossilization provides much information about the animal under study.

Several specimens of the lungfish *Scaumenacia curta* are preserved in three dimensions, revealing the mobility and flexibility of the animal's body. Above: two different views of the same fossil, showing the fins in a life-like position, attached to the slightly curved body of the fish.

Photos : APM
Illustrations : R. Cloutier

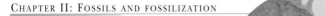

Cross-section of a *Eusthenopteron foordi* tooth, observed in polarized light. One can see the "labyrinth" structure of the dentine and also (blow-up) the organization of the dentine fibres that make up this tissue. The left-hand drawing depicts a tooth from an osteolepiform closely related to *Eusthenopteron*.

Photos : Moya M. Smith, United and Medical Dental School, Guy's Hospital, London. Illustration : M. Liepina, Stockholm.

Furthermore, even microscopic details can persist after 370 million years. It is possible to identify various histological structures, namely the tissues of mineralized body parts. This is the case with dentine and enamel, two of the main tissues that still make up the teeth of tetrapods today, including humans. Cutting through fossil fish teeth reveals fine structures that are visible under a microscope.

The overwhelming majority of fossils tell us about the hard, mineralized parts of organisms. The soft parts, i.e., vital organs such as the heart, the stomach, the intestines and the brain, or the muscles and fatty tissues, usually decompose shortly after death, well before any actual fossilization takes place.

Preceding page, below: this specimen, composed of several thousands of scales and bones, shows the details of the head, the hind section of the body, and the pectoral fins of the acanthodian *Homalacanthus concinnus*.

Photos : APM

Occasionally, however, these soft parts are preserved. For this to happen, exceptional conditions are needed for fossilization. This was the case at Miguasha 370 million years ago. Today, a large number of its fossil fish specimens display traces of soft parts. Gill imprints and certain eye parts have, for example, been found in certain fish from the site. Finally, Miguasha has recently been the scene of an important discovery. Traces of blood vessels have been detected in a specimen of an armoured fish. The finding is a scientific first and confirms the quality and importance of this fossil site.

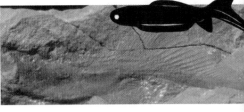

*E*ndeiolepis aneri, a jawless fish with no ossified parts.

The fine-grained silts of the Miguasha cliffs, however, have preserved several complete specimens. Above: anal fin and tail of the anaspid *Endeiolepis*.

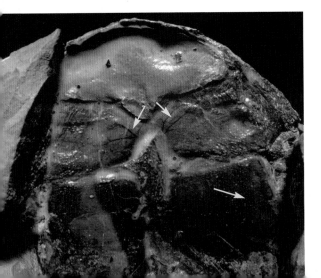

*V*entral view of the hind section of a placoderm *Bothriolepis canadensis*. The outline of blood vessels is visible. Fossilization of the blood vessels is believed to be due to oxidation of the blood, which coloured the fine-grained sediments in which the animal was preserved.

Photos : APM

CHAPTER III

LIFE AND THE EARTH

The Earth is a special heavenly body. The presence of liquid water on its surface has allowed life to develop.

Photo
JSC
NASA

Life appeared very early in the Earth's history, at least 3.5 billion years ago—about a billion years after our planet was formed. For nearly 3 billion years, almost the only living things were bacteria and other one-celled life forms. Not until 600 million years ago did an explosion occur in biological diversity, a sort of "big bang" of life.

But how did it all start ?

EVOLUTION

THE THEORY OF CHARLES DARWIN

Georges Louis Leclerc, comte de Buffon (1701-1788) supervised the writing of the *Histoire naturelle générale et particulière*, a 44-volume work that has never been equalled in its magnitude and impact on the public. It restructured contemporary views on biological diversity.

The chevalier de Lamarck (1744-1829) coined the word biology for the science of living things and is considered to be the founder of this discipline. He also originated the word ecology, the science of the relations between living things and their environment.

Until the 19th century, the dominant view in biology reflected popular and religious beliefs. People believed that each species had been created in an immutable form when life began and had kept that form ever since.

Over time, many enlightened thinkers gradually altered this view of life's origins. An 18th century French naturalist, Georges-Louis Leclerc, Comte de Buffon, was the first to argue that the Earth was much older than the 6,000 years in the Bible. Others who helped change the accepted history of life included the Frenchmen Jean-Baptiste-Pierre Lamarck, Pierre-Louis Moreau de Maupertuis,

and Georges Cuvier—the founder of comparative anatomy, who made vertebrate paleontology a true science—and the British scholar Erasmus Darwin. It was the grandson of Erasmus, the British naturalist Charles Darwin, who in the 1850s laid the foundations for the now famous theory of evolution. Charles Darwin explained the evolution of species by the action of natural selection. According to his theory, random variations occur in certain characteristics. If beneficial, individuals who have them will stand a greater chance of surviving and passing them on to their descendants. If not, the variations will be eliminated. Thus, animals and plants will slowly change from one generation to the next, becoming ever more adapted to their environment.

To back up his theory, Darwin amassed a wide range of evidence. He had, for example, noticed much similarity between fossil and living species within the same geographical area and he deduced that "kinship" existed between them. After observing many bird species at close hand, differing from each other in a few characteristics, he reasoned that they all derived from a common ancestor. To explain these varieties of closely related species, Darwin drew a parallel with livestock breeders who produce new varieties, sometimes quite different from

At the age of 22, Charles Darwin (1809-1882) made a 5-year voyage aboard the vessel the Beagle. In the course of this journey, he amassed a considerable amount of geological and biological observations, with which he went on to develop his theory.

This caricature of Darwin (left) reflects the controversy his theory stirred up in the 19th century and much of the 20th.

each other, by continually selecting the animals they would allow to breed.

Darwin's theory was long criticized, being totally at odds with prevalent Christian beliefs in the late 19th century. Today, it is widely accepted throughout the scientific community. Some refinements have admittedly been added since its publication in 1859, particularly as a result of discoveries in genetics.

A MORE MODERN VIEW OF EVOLUTION

The now famous double-helix model of the DNA molecule. Its structure was discovered by James Watson and Francis Crick in 1953, earning them the Nobel Prize, nine years later. All of the genetic information that determines a human's makeup is contained in the "bars" of this ladder, which is present in every cell.

Genetics explains the mechanisms of evolution through changes in genes, called mutations, and through changing proportions of different versions of genes, called alleles, in a population. Alleles that benefit the organisms that have them are more likely to be passed on to future generations, thus becoming more common. Bit by bit, individuals with beneficial alleles will come to replace individuals with less beneficial ones. The latter are weeded out by natural selection.

Since the early 1980s, researchers have worked hard at retracing kinship ties between species by looking for traits that are specific to different groups. Their work has allowed us to reconstruct the great family tree, or phylogeny, linking extinct and contemporary life forms to each other.

It has also helped us deduce the evolutionary mechanisms that generated these kinship ties. This discipline is called "cladistics" and is now fundamental to evolutionary studies. Over a century after Darwin, we are piecing together over time and space the long evolutionary chains of fossils and living species, thus forming a hierarchy of forms in which the living ones are the topmost links of the chains.

In the short and medium term, the survival of individuals is explained by the fundamental principles of reproduction, survival, and natural selection. Evolution, however, is also about changes that happen over many thousands or even millions of years. Here, two additional notions come into play: the appearance of new species, called speciation, and the disappearance of existing species, called extinction. The magnitude of our current biodiversity is the most eloquent testimony to the speciation that has occurred over the history of the Earth.

HOW DOES A NEW SPECIES APPEAR?

For the past fifty years, population biology and evolutionary biology have been striving to understand the conditions and mechanisms for the appearance of species. A new species appears by branching off from an ancestral one. For example, a geographical barrier may separate individuals from the same species into two groups, or populations. The barrier usually results from a geological event and will keep individuals in one population from meeting individuals in the other. The formation of a mountain chain and the separation of land by ocean are two striking examples. Once isolated, the two populations must diverge from each other in some characteristic, be it molecular, anatomical, behavioural, or physiological. The difference, whatever it is, originates in a genetic

Photo : Gérard Massé - FAPAQ

The copper red-horse (*Moxostoma hubbsi*) is the only Quebec vertebrate species whose world geographical range is confined to a few rivers in the Montreal area. It is also the only Quebec fish species that has been declared in danger of extinction. Currently, only a few hundred are left in the wild.

mutation in an individual's reproductive cells. The modified DNA is passed on to the individual's descendants and, if beneficial, will spread through that particular population. Genetic mutations may have many sources, such as exposure to hazardous chemical and physical agents (e.g., the sun's ultraviolet rays) or malfunctions in the cellular machinery that assembles the molecules of living things. These mutation sources help maintain variability within populations. In conjunction with other factors, notably ecological ones, such mutations can cause new species to appear.

Over time, in response to climate change, to competition between species, or to an isolated, random and often catastrophic event, e.g., a meteorite impact or a period of exceptional volcanic activity, a species will gradually change and adapt to its new environment. Or it will go extinct.

WHAT ARE THE EVIDENCES FOR EVOLUTION?

What events make us realize that evolution exists? Over a human lifespan, it is certainly hard to see animals or plants evolve. Evolution usually results from slow natural processes that eventually become perceptible over many generations. If we look a bit closer, though, we can easily discover much evidence of evolution around ourselves, particularly in biological diversity, in the apparent order of living creatures, and in the richness of the fossil record.

Today, there are believed to be 30 million living species—the total depends on the way they are counted. This biodiversity, and the geographical range of these species, is evidence for evolution. Each living animal or plant form is adapted to its environment, with feeding behaviours and other responses specific to its surroundings. Chance alone is not a satisfactory explanation. Nor are independent acts of creation for each and every species. Much more satisfactory is the hypothesis of evolution from common ancestors that appeared in specific locations and gradually adapted to these locations.

Reconstitution of a major meteorite impact on the Earth's surface.   It was the probable cause of a mass extinction 65 million years ago.

Illustration : Don Davis - NASA

NUMBER OF SPECIES

EUBACTERIA + ARCHAE	4,000
PROTOCTISTA	104,000
Including :	
Actinopods	6,000
Foraminiferans	34,000
Ciliates	8,000
Sporozoans	5,000
Rhodophytes (red algae)	5,000
Gmophytes (green algae)	10,000
Bacillariophytes (diatoms)	12,000
PLANTS	270,000
Including :	
Bryophytes	16,000
Pteridophytes	10,000
Spermatophytes	240,000
FUNGI	70,000
ANIMALS	1,400,000
Including :	
Sponges	10,000
Cnidarians	10,000
Flat worms	25,000
Nematodes	80,000
Arthropods	1,085,000
Including :	
Crustaceans	40,000
Arachnids	75,000
Insects	950,000
Including :	
Beetles	400,000
Flies	120,000
Hymenopterans	130,000
Lepidopterans	150,000
Mollusks	100,000
Annelids	12,000
Echinoderms	6,000
Chordates	54,000
Including :	
Fish	29,000
Amphibians	4,000
Reptiles	6,500
Birds	9,67?
Mammals	4,327
Including :	
Rodents	1,702
Primates	23?

Out of an estimated 30 million animal and plant species now in existence, fewer than 2 million have been identified to date, 950,000 being insect species! For all of these species, however, there is an order, a hierarchy of life forms. It is possible not only to classify species into groups of related life forms, but also to create a classification based on their common anatomical and molecular characteristics. This hierarchy covers all kingdoms of living things, from the simplest to the most complex. Over the past ten years, molecular geneticists and systematists have discovered long DNA sequences that are identical in both bacteria and humans.

Previous page: a statistical overview of our biodiversity. The immense variety of species is proof of evolution.
Quebec alone has all of the species in the figure on page 54 and is home to a sample of most of the groups named, despite its northerly latitude.

Illustrations :
R. Cloutier

Such molecular similarities within the actual messages of our genes could not be pure chance. They are further proof of evolution.

This hierarchy of life forms appears not only at the molecular level but also in embryonic development. It is almost troubling to see the strong similarities that exist between distantly related animal species, such as a fish and a mammal, during the initial stages of their embryonic develop-

A simplified phylogenetic tree showing the hierarchy of groups and also the morphological characters that are unique to the representatives of each group.
Left to right: outlines of the osteolepidida *Osteolepis*, the eusarcopterygian *Panderichthys* and the Devonian tetrapod *Acanthostega*.

ment. Even without looking at embryos, we need only compare the anatomy of animals to see that the same "building plan" has been followed in species with very different ways of life. Vertebrate forelimbs are a case in point. The arm of a human, the fin of a whale, and the leg of an early land tetrapod 360 million years ago, for example, are astonishingly alike. The skeletal resemblances exist because these different species must have had a common ancestor, from which they have subsequently evolved.

RECONSTRUCTING THE TREE OF LIFE

When tracing your family's origins, you may draw a genealogical tree and write the names of your ascendants and descendants on the branches, and the kinship ties between them. Phylogeny is for groups of organisms what genealogy is for a single individual. A phylogenetic tree displays the evolutionary relationships between species or different groups of plants and animals. These phylogenetic trees, or dendrograms, are the fruit of work in cladistics, a discipline developed by the German entomologist Willi Hennig in 1950. Through detailed study of anatomical or molecular traits, one can determine how evolutionary innovations are distributed and then trace phylogenetic relationships. For example, the presence of fingers is an evolutionary innovation that unites all tetrapods.

In the background can be seen an evolutionary tree drawn in the late 19th century. Almost a work of art...

The History of Creation, vol. II, (1876), Ernst Haeckel. D. Appleton and Co., New York.

About thirty million living species make up our current biodiversity. For many millions of years, though, species have been coming into existence, persisting through time, succeeding other species, and dying out. Since life began on earth, over 3.5 billion years ago, an immense number of species have lived on our planet. The outline of this past can be pieced together from telltale evidence left behind by the passage of time and geology, that is, from fossils.

This fossil record is made up of the fossil species found so far and represents but a tiny proportion of all species that have inhabited the earth. Links are missing from the long chain that connects together the living and fossil species described to date. Until the 1960s, biologists and paleontologists considered the fossil record to be the only inescapable proof of evolution. Although other kinds of proof are available today, fossils are nonetheless the only traces we have of the presence of life on earth millions of years ago. The study of fossils provides us with the information we need to reconstruct life at Miguasha in the Devonian.

A superb example of the quality of preservation at Miguasha. Two specimens of the lungfish *Scaumenacia* and two specimens of the placoderm *Bothriolepis* on the same slab.

The above diagram outlines the positions of the four fossils on the slab

Photo : APM
Illustration : R. Cloutier

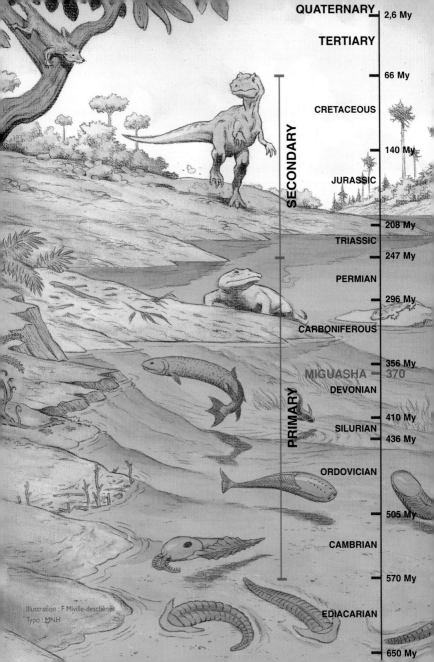

QUATERNARY — 2,6 My

TERTIARY — 66 My

SECONDARY

CRETACEOUS

— 140 My

JURASSIC

— 208 My

TRIASSIC — 247 My

PERMIAN

— 296 My

CARBONIFEROUS

— 356 My

MIGUASHA — 370

PRIMARY

DEVONIAN

— 410 My

SILURIAN — 436 My

ORDOVICIAN

— 505 My

CAMBRIAN

— 570 My

EDIACARIAN

— 650 My

Illustration : F. Miville-deschênes
Typo : MNH

EVOLUTION THROUGH THE EYES OF PALEONTOLOGY

THE GEOLOGICAL TIMESCALE

Life appeared very early in the Earth's history, at least 3.5 billion years ago. For nearly 3 billion years, almost the only living things were bacteria and other one-celled life forms. Not until 600 million years ago did an explosion occur in biological diversity, a sort of "big bang" of life. Thus began the first era of visible life, or the Paleozoic, from 570 million to 250 million years ago. The results of this explosion of biodiversity some 600 million years ago live and grow every day under our eyes.

THE ORIGIN OF LIFE

Until the 18th century, the only widely accepted theory of the origin of life was the theory of spontaneous generation, which had the advantage of not challenging the Genesis account too seriously. The same belief crops up in the ancient writings of China, India, and Egypt. According to the Greek philosopher Aristotle, "plants, insects, and animals can arise from living systems that resemble them, but also from decomposing matter activated by the heat of the sun." This belief was finally laid to rest by the French scientist Louis Pasteur, a contemporary of Charles Darwin.

The geological timescale (previous page) brings together some inhabitants of our planet that have succeeded each other over the past 650 million years. From bottom to top: *Spriggina* (Ediacara fauna, identified as either an annelid or an arthropod), *Anomalocaris* (Burgess Shale fauna), the primitive vertebrate *Sacabambaspis*, the osteolepiform *Eusthenopteron* and the placoderm *Bothriolepis* (Miguasha), the amphibians *Eryops* and *Trimerorhachis*, a dinosaur *Allosaurus*, and a primitive mammal *Morganucodon*.

Aristotle (384-322 B.C.)

In 1860, Pasteur developed a reliable method of sterilization and demonstrated that life could not arise spontaneously

from inanimate matter. His discovery, incidentally of great use for medicine, put an end to the theory of spontaneous generation.

After the work of Pasteur and Darwin, evolutionary thinking was quickly extended to inert matter. In the first half of the 20th century, scientists like the Soviet theorist A.I. Oparin and the British thinker J.B.S. Haldane linked the origin of life to the formation of the Earth. They proposed that the different molecules essential to life—the "building blocks" that make up the cells of a living being—were initially produced by a wide variety of chemical reactions that occurred 4.5 billion years ago when the Earth's atmosphere was created. In the 1950s, a young American student, Stanley L. Miller, reconstituted these "primitive" conditions in a laboratory. Using simple components that should have been present in the early atmosphere, Miller, managed to obtain more complex organic molecules, including certain amino acids — the building blocks of proteins.

Louis Pasteur (1822-1895; above at 30) made decisive discoveries in chemistry, human and veterinary medicine, agriculture, and food processing. He is still famous for having developed the rabies vaccine in 1885.

Photo : Bibliothèque Nationale.

The strength of these experiments lies in their demonstration that such "basic building blocks of life" could be made

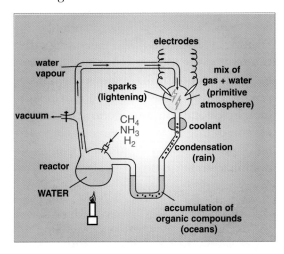

Stanley Miller's experiment: A vacuum is created in a closed sterile reactor. Three gases (methane, CH_4; ammonia, NH_3; and hydrogen, H_2) are introduced and mix with water vapour, thus recreating the primitive atmosphere. Organic molecules then form in response to energy from electrical sparks that simulate lightning. The molecules become concentrated in "oceans" of condensed water vapour.

through chemical reactions in natural environments. It was a big step forward in providing new leads back in time not just to the first organism but also to the first molecules that may have formed the first organism. Nevertheless, not all questions on the origin of life have been resolved and, even today, scientists are divided over the initial scenario for the appearance of life on Earth. Whatever the scenario, it is now accepted that the appearance of life was initially followed by a very slow evolution over three billion years, until the explosion of biodiversity—about which the fossil record provides us with a few secrets.

Illustration : P.-A. Bourque, Univ. Laval.

THE EDIACARA FOSSILS: ENIGMAS OF THE PRECAMBRIAN

Stromatolites from Lake Mistassini, Quebec, going back 1.8 billion years. A two-coloured metre stick lying on top indicates the size of these Precambrian fossils. Below: a cross-section of stromatolites showing a succession of sedimentary layers.

The quest for traces of the first organisms to inhabit the Earth began about a hundred years ago and led paleontologists to examine the world's oldest sedimentary rocks and discover fossils that scarcely resemble living things. These are stromatolites: limestone columns made up of a succession of fine layers of sediments deposited by photosynthetic bacteria, known as cyanobacteria.

These cyanobacteria, which still exist, are very primitive one-celled organisms that can carry out photosynthesis, i.e., the use of energy from the sun's rays to make the molecules needed for growth, as in plants. Today, stromatolites may be observed on Australia's Great Barrier Reef and it is precisely Australia that has yielded the oldest stromatolites, dating to 3.46 billion years ago. In Quebec,

10 cm

Photos : Hans J. Hofmann, McGill Univ.

Nunavut, and Ontario, assemblages of microfossils have been found and dated to almost 2 billion years ago.

For three billion years, the only life forms inhabiting the Earth were one-celled organisms, mainly bacteria. The first traces of many-celled organisms, or metazoans, date back about a billion years and take the form of a few ichnofossils. The development of life then speeded up 600 million years ago in the late Precambrian, before the Paleozoic era. Fossils of strange fauna have been discovered, initially in the Ediacara Hills of Australia and later at about twenty other sites spread over five continents, including sites in Western Canada.

These organisms were mostly soft-bodied and are hard to assign to the animal and plant groups we know today. Understanding this primitive fauna is also complicated by a lack of any mineralized parts. They have left only their imprints, leaving much room for interpretation by the paleontologist and thus retaining much of their mystery.

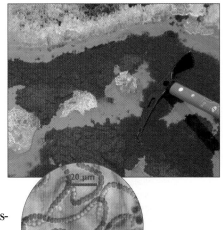

An example of an extreme environment that cyanobacteria today live in: a carpet of cyanobacteria found in an icecap.

Blow-up of *Nostoc* cyanobacteria 20 microns (μm) or 20 thousandths of a mm across.

Photos : Warwick Vincent, Univ. Laval

The Ediacara fauna nonetheless bear witness to a major step in evolution, being the first assemblage of metazoans to have come to our attention. Most of these organisms soon disappeared some fifty million years later —a mere instant in geological time.

THE EXPLOSION OF LIFE

The Ediacara Hills of Southern Australia whose fossils were first discovered in 1947. They include *Mawsonites* (left), a species of jellyfish around 13 cm in diameter, and *Dickinsonia* (right), a flat annelid worm 9 cm long.
Below, *Parvancorina*, a sort of small crustacean 2.5 cm long and unlike any known animal.

Photos : South Australian Museum (Ben McHenry, Dir., Collections Ediacara).

The appearance of metazoans a billion years ago, and their diversification and dispersal around the world 600 million years ago, was followed by another big step: the mineralization of the external skeletons of certain living things, e.g., shells and carapaces. The real explosion, however, in terms of biological diversity and complexity, came in the Middle Cambrian around 530 million years ago. Proof of this event was furnished in the early 20th century by the discovery of the Burgess Shale fossils at Mount Stephen in British Columbia, followed by the discovery of other sites around the world. The Burgess Shale site has yielded specimens of 120 different kinds of organisms.

This total represents an enormous diversity, all the more diverse given that only about 30 of the 120 belong to phyla (divisions of the animal kingdom) that are well known to scientists. So, like the Ediacara fauna, a large part of this fauna must have gone extinct without leaving any descendants. All of these specimens were, so to speak, testing out the anatomical organization of living things and only a minority have lasted over evolutionary time. Most of the modern divisions of the animal kingdom, and a wide array of animal forms that soon died out, appeared over a short period 530 million years ago.

The main characteristic that these organisms had in common was an external skeleton. Whether it was the calcium carbonate shell of the mollusks or the chitinous casing of the

The Burgess Shales. After being falsely attributed to shrimps, it has been shown that the prehensile appendages shown below actually belonged to *Anomalocaris*, a Cambrian carnivore that could reach 60 cm in length and was apparently made up of mostly of soft tissue..

Photos : Andrew MacRae

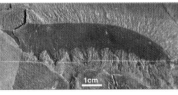

1cm

arthropods, these animals had developed a system to protect themselves from possible predators and from the environment they lived in.

Such a "big bang" of life forms has never reoccurred. Almost all organisms on earth today are variants of basic models established back then in geological time.

The First Vertebrates: Who Are Our Ancestors?

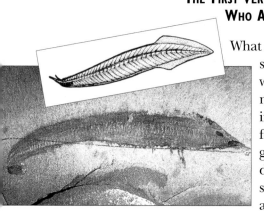

What do sea urchins and starfish have in common with humans? Much more than you might imagine! Of all living or fossil invertebrate groups, the echinoderms—sea urchins, starfish, and sea lilies— are the ones that share the most characteristics with humans and with vertebrates in general—fish, amphibians, reptiles, birds, and mammals. The evolution that led from echinoderms to vertebrates took place gradually, through different intermediary stages. We see this anatomical progression, mainly in the nervous system, in three groups of aquatic animals still living today: hemichordates, urochordates, and cephalochordates.

Pikaia gracilens (fossil and reconstitution shown above) is considered to be the first chordate, i.e., an animal with a supporting rod running down its back. It belongs to the Burgess fauna described previously.

Photo : Simon Conway-Morris, Cambridge University

The distant common ancestor of echinoderms and vertebrates unfortunately remains unknown. The fossil record does, however, tell us about the first animals with the beginnings of a backbone—the characteristic structure of vertebrates.

In the 1980s, a Quebec paleontologist, Pierre-Yves Gagnier, discovered in the Andean highlands of Bolivia the remains of one of the oldest vertebrates on earth. It was *Sacabambaspis*, a fish that lived 475 million years ago, or around 100 million years before the remains found at the Miguasha cliffs. About 30 cm long, its spindle-shaped body had a tail, long narrow scales, and a head covered with large bony plates.

An almost complete specimen of *Sacabambaspis janvieri*, a primitive vertebrate discovered in Bolivia.

Sacabambaspis had tiny eyes in the front of its head, a small mouth probably for filtering mud and, on both sides, a series of small plates protecting the gills it used for respiration.

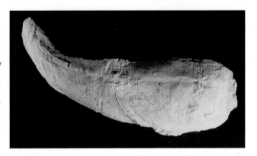

Photo : APM

The Bolivian fossils have filled a gap in paleontology. Before this discovery, the first vertebrates were known only from tiny scales, less than a millimetre in size, or from a few rare fossils showing just a small part of the organisms. Such microfossils from the Middle Ordovician, around 450 million years ago, have been found in the United States, Australia, eastern Siberia, and also the limestone deposits of the St. Lawrence Lowlands and the Charlevoix region. Although no complete specimen has yet been found in Quebec, such a discovery may one day unravel the mystery of the first vertebrates.

In southern China, fieldwork carried out in 1999 may push back the origin of vertebrates. The Chengjiang site contains many fossils of two species that resemble very primitive fish.

These 2 to 3 centimetre-long animals date from the Early Cambrian, 540 million years ago, and have specifically vertebrate characteristics according to some paleontologists. If they prove to be the oldest vertebrates found to date, our division of the animal kingdom would have appeared some 70 million years earlier than scientists had previously thought—at the time of the "big bang" of life described in the last chapter.

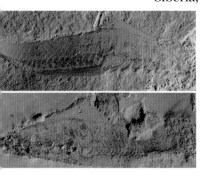

These specimens of two vertebrate species were discovered in Chengjiang (*Myllokunmingia* above and *Haikouichthys* below). They seem quite close to the reconstitution of what the first vertebrate looked like, as proposed by Holmgren and Stensiö in 1937

Photos : Degan Shu, Xi'an Univ.
Illustration : Holmgren et
Stensiö. – Philippe Janvier
(2000), Les premiers poissons
pris dans les filets chinois ;
La Recherche, vol. 329, p. 18-19.

THE CONQUEST OF DRY LAND

Everything leads us to believe that small millipedes were the first invertebrates to colonize the dry land, around 450 million years ago. A wide array of small arthropods, including mites and spiders, followed them over the next 50 million years. Under a protective cover of vegetation, there slowly developed a complete ecosystem: a fertile land environment teeming with microscopic life and diverse invertebrates. Such an ecosystem was a prerequisite for a key step in evolution: the conquest of dry land by vertebrates.

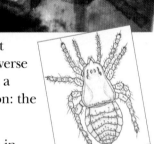

Not long after their appearance in the waters of the globe, the vertebrates greatly diversified. Several groups of fishes succeeded each other over a period of around 200 million years, the first ones being jawless. Finally, in the Devonian, some of them had evolved to the point that they could timidly move out onto the dry land, already far from barren.

The trigonotarbids, from the Rhynie fossil beds in Scotland, are the oldest known land animals. These arthropods, of about one centimetre in size and resembling spiders, appeared 400 million years ago and died out 150 million years later.

Photo & diag.: William A. Shear, Hampden-Sydney College, USA

The marine vertebrates had to face several formidable problems before colonizing the land surface. Acquiring a means of locomotion appropriate to dry land, using the oxygen in the air to breathe, and reproducing successfully out of water were not the least of these problems. The resulting morphological and physiological changes came about slowly over many millions of years. It was a case of adapting to a new environment—an evolutionary challenge.

Footprints of a Devonian tetrapod from Australia.

Photo : Muséum national d'Histoire naturelle de Paris

Some fishes found at Miguasha, such as the lobe-finned fishes *Eusthenopteron foordi* or *Elpistotege watsoni,* to be discussed further on, show evidence that these animals had begun an evolution toward "objective Land."

The exceptional conditions of fossilization at Miguasha made it possible for this *Eusthenopteron foordi* to be preserved in a curved position.

Photo : APM

Among the tetrapods, amphibians have remained dependant on an aquatic environment, in particular for reproduction. This is where they generally lay their eggs, thus limiting their habitats to areas near water. Colonizing areas further inland may have been made possible by a remarkable innovation: an egg inside a calcareous shell and an amniotic sac—a membrane that encloses the fetus and protects its development. For this reason, tetrapods that have an amniotic sac during embryonic growth are called amniotes.

Complete specimen of *Acanthostega gunnari*, one of the oldest known tetrapods (360 million years old) discovered in east Greenland.

Photo : Jennifer A. Clack, University Museum of Zoology Cambridge.

The first amniotes were reptiles. They could lay their eggs on dry land, unlike amphibians, and begin to move into ecological niches far from water. Highly adaptive, they soon dominated the interior of each continent. Then, many of them went extinct in the Upper Cretaceous, including the famous dinosaurs.

The blue-spotted salamander (*Ambystoma laterale*) occurs in Quebec. This amphibian lays its eggs one by one at the base of plants or rocks

Photo : Pierre Etcheverry

Elpistostege
watsoni

Homalacanthus
concinnus

Plourdosteus
canadensis

Cheirolepis
canadensis

Miguashaia
bureaui

Eusthenopteron
foordi

Scaumenacia
curta

Triazeugacanthus
affinis

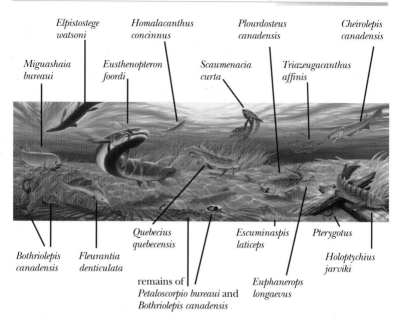

Bothriolepis
canadensis

Fleurantia
denticulata

Quebecius
quebecensis

Escuminaspis
laticeps

Pterygotus

Holoptychius
jarviki

remains of
Petaloscorpio bureaui and
Bothriolepis canadensis

Euphanerops
longaevus

Illustration : F. Miville-Deschênes
Photo : APM

Scaumenacia
curta

Eusthenopteron
foordi

Elpistostege
watsoni

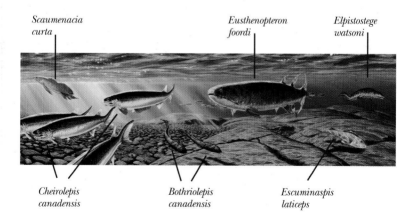

Cheirolepis
canadensis

Bothriolepis
canadensis

Escuminaspis
laticeps

Illustration : Robert J. Barber
Photo : Dennis Finnin, AMNH

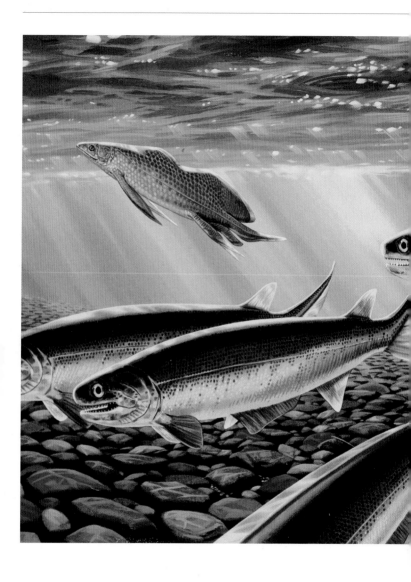

Illustration : Robert J. Barber
Photo : Dennis Finnin, AMNH

Illustration : F. Miville-Deschênes
Photo : APM

THE AGE OF DINOSAURS, LARGE AND SMALL

Though rather small, the first reptiles were to give rise to such creatures as the great saurians, known as dinosaurs. Reptiles ruled the world during the Mesozoic, from 250 to 66 million years ago, colonizing most of the available ecological niches. Dinosaurs occupied the land, pterosaurs and pterodactyls the air, and ichthyosaurs, plesiosaurs, and mosasaurs the water. These reptiles measured 15 centimetres to 15 metres long, weighed 100 grams to 20 tons, were either bipeds or quadrupeds, and were either carnivores or herbivores.

Protoceratops andrewsi: this herbivorous dinosaur from Mongolia lived in the Upper Cretaceous 75 million years ago. As an adult, it could reach two metres in length.

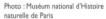

Photo : Muséum national d'Histoire naturelle de Paris

Then, around 66 million years ago, the dinosaurs disappeared. The exact cause still eludes us although there was probably more than one. Among the causes put forward, scientists now favour a collision between the Earth and an asteroid estimated at 10 km in diameter. Several pieces of evidence back this

Nine meteorite craters have been identified in Quebec. They range in size from 3.2 km (New Quebec crater, left) to 88 km (Manicouagan, page 74).

Photo : Daniel Lamothe (Jehan Rondot, 1995, *Les impacts météoritiques à l'exemple de ceux du Québec* ; MNH, Québec.

METEORITE IMPACTS AND QUEBEC

Each year, tens of thousands of tons of extraterrestrial material fall through our atmosphere. Of this quantity of rock, only a few meteorites leave any traces and even fewer produce large craters, also called astroblemes. On average, a large asteroid strikes the Earth every 100,000 years. According to geologist Jehan Rondot, nine craters of meteorite origin have been identified in Quebec, craters ranging in size from 3.2 km (New Quebec Crater) to 88 km (Manicouagan). Two of these meteorite impacts occurred during the Devonian—one in Charlevoix and the other at Lac La Moinerie (near Ungava Bay). The size of the Chixculub Crater, linked to the great extinction of the Upper Cretaceous, is around 200 km. Some suggest that the environmental effects of the Charlevoix astrobleme, 56 km across, may have contributed to the extinction of many species in the Upper Devonian. [1]

O n this mosaic of radar images of the Manicouagan reservoir are indicated the diameter of the astrobleme (large circle) and the central upraise (small circle).

1 Rondot, J. 1995. *Les impacts météoritiques à l'exemple de ceux du Québec.* MNH, Quebec City.

hypothesis, in particular the Chixculub crater, 200 km across, on the Yucatan Peninsula in Mexico and a layer of disturbed sediments from the impact, ranging from a few centimetres in depth to a few metres, discovered at different locations around the world.

The collision between this asteroid and the Earth cannot, alone, explain the extinction of the dinosaurs. Such a shock would probably have caused an enormous tidal wave and the expulsion of a phenomenal amount of dust into the atmosphere, thereby forming a solar shield that limited plant photosynthesis and thus the production of oxygen. It is this set of causes that is now linked to the extinction of the dinosaurs.

"WHERE HAVE THE DINOSAURS GONE?" *

First, we should keep in mind that 75% of all species inhabiting the world's surface disappeared at the same time as the dinosaurs did in the Upper Cretaceous. Such a holocaust is called a mass extinction. Many others have occurred during the earth's history and the one that killed off the dinosaurs some 66 million years ago was not the worst. Although most dinosaurs of the Upper Cretaceous died out, the direct descendants of one group of small feathered dinosaurs are still with us. We call them birds.

(p.74) Photo : NASA. (Jehan Rondot (1995), *Les impacts météoritiques à l'exemple de ceux du Québec*, MNH, Québec).

Photo : Pierre Etcheverry

THE WAY IS CLEAR FOR MAMMALS

Appearing around 225 million years ago, mammals derive from a special group of amniotes. Initially the size of shrews and insectivorous, they lived in the shadow of the great reptiles until the mass extinction of the Upper Cretaceous. The mammals took advantage of the numerous vacated habitats and ecological niches, rapidly expanding and diversifying to form many species during the Tertiary period. By 50 million years ago, most of the current

The smoky shrew is found in Quebec and feeds on worms and insects. Shrews are considered to be living forms of primitive mammals.

* Like Quebec singer Michel Rivard, it is legitimate to ask: "Where have the dinosaurs gone?"!

orders of mammals were already on the earth: chiropterans (bats), cetaceans (whales), proboscideans (elephants), and carnivores, to name just a few. Primates go back some 65 million years, hominids 15 million years, and modern humans, *Homo sapiens*, less than 100,000 years.

Our history is quite short when compared to the 4.6 billion years of the planet Earth! And our planet, too, has evolved and changed since its formation.

OUR CHANGING PLANET

The Earth's crust rises and falls. Its oceans open and close. Its continents form and erode. All of these changes that shape our planet are, over a human lifespan, infinitely slow. To grasp them, we can turn to the different organisms preserved as fossils, which have also undergone these changes. Fossils help us piece together what the earth looked like many millions of years ago.

CONTINENTAL DRIFT

When the Canadian paleontologist Whiteaves described the Miguasha fishes in the 1880s, he noticed similarities between this new Devonian fossil fauna and the fauna discovered by Hugh Miller, some 40 years earlier in the sedimentary Devonian rocks of Scotland. The resemblances between the Quebec fish remains and the

Volcanic eruptions are one of the most impressive manifestations of geological activity on our planet. Above: Mount St. Helens, volcano in the north-western U.S., whose first historical eruption took place on May 18, 1980.

Photo : Austin Post, US Geological Survey / Cascades Volcano Observatory

Scottish ones intrigued the Canadian paleontologist. Why were the fossils so similar?
They dated to the same period but came
from two sites thousands of kilometres
apart. The question remained unanswered
until the 1960s.

As early as the 17th century, when people still believed that the Earth was less than
6,000 years old and created in six
days, some observers, such as the
English philosopher and scientist
Francis Bacon, noticed that the east
coast of South America and the
west coast of Africa could fit into
each other like pieces in a giant
puzzle. In the 19th century, geological resemblances between the two
continents were discovered, but a
genuine understanding of these
trans-Atlantic similarities had to await the
new ideas of the German geophysicist and
meteorologist Alfred Wegener, in the early
20th century. In 1915, Wegener used geological, paleontological, and paleoclimatic
data to propose that, 250 million years ago,
all of the current continents were part of a
single "supercontinent" called Pangea, a
Greek word meaning "all of the Earth."
Pangea then broke up into several pieces
that slowly drifted apart.

For forty years, Wegener's contempo-
raries rejected his theory, very novel at the
time. With the development of sonar during

Alfred Wegener
(1880-1930)
below in 1930.
He was conducting
an expedition to
Greenland when he
died a few days after
his fiftieth birthday.

Photo : J. Georgi
Archives Alfred-Wegener-
Institut für Polar- und
Meeresforschung

the two world wars, studies of the Atlantic seafloor in the 1960s confirmed the continental drift theory. These results, together with data from paleontology, paleomagnet-

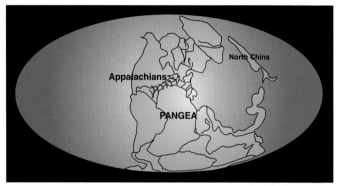

A map of Pangea in the Upper Permian 255 million years ago.

Illustration : C.R. Scotese (1997) Paleogeographic Atlas, PALEOMAP progress report 90-0497, Department of Geology, University of Texas at Arlington.

The diagram (opposite page, right) shows the movement of tectonic plates, i.e., the external layer of the earth, comprising the crust and upper mantle.

ism, volcanology, and seismology, have provided the model widely accepted today in our understanding of all geological phenomena.

The Earth's surface, including the continents and the seafloors, is made up of a rigid crust less than 100 km thick and divided into a dozen moving plates, which carry and move the continents. The "engine" at the origin of these movements is the convection that takes place within the earth's mantle—a relatively plastic environment on which the crust rests. There are thus places on the earth—the mid-oceanic ridges—where new crust is continually being created and other places—the subduction zones—where old crust is being destroyed. Between the mid-oceanic ridges and the subduction

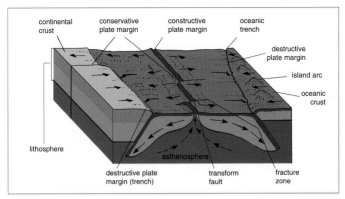

continental crust — conservative plate margin — constructive plate margin — oceanic trench — destructive plate margin — island arc — oceanic crust — lithosphere — asthenosphere — destructive plate margin (trench) — transform fault — fracture zone

zones, the plates move at a speed of 2 to 10 cm per year. Over many millions of years, they bump into each other, move away from each other, or grind against each other. This bumping and grinding causes intense "tectonic" geological activity that can lead to earthquakes, volcanic eruptions, and mountain building. Such tectonic activity is still producing effects today!

This model, called plate tectonics, has had a revolutionary impact on earth science. It has helped us understand the distribution of both fossil and living species over the earth's surface by explaining the formation of geographical barriers.[1] Using this model, we can deduce Miguasha's geographic location and its surrounding landscape 370 million years ago.[2]

Illustration : *The Ocean Basins : Their Structure and Evolution* (1989). The Open University, Milton Keynes.

The San Andreas Fault in California is a transform fault between the Pacific and North American plates. The term refers to a fault where two plates slide against each other without creating or destroying crust, although these regions are prone to earthquakes.

Photo : Robert E. Wallace, US Geological Survey

[1] see p. 51
[2] see p. 100

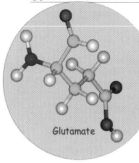

Glutamate

We have here the structures of three of the most abundant amino acids in the meteorite. Green: carbon atoms. Blue: nitrogen atoms. Red: oxygen atoms. White: hydrogen atoms.

DID LIFE ARISE ON EARTH OR WAS THE EARTH "SEEDED"?

Until very recently, life and all of the complex building blocks of life were known only on our planet. So the question never arose. Life was clearly of earthly origin.

Then, in 1970, a team of biologists, astronomers, and geologists examined a meteorite and found certain molecules that are basic to all living

Alanine

things: amino acids. It now became possible to imagine an extraterrestrial origin for life, or at least for its building blocks. A meteorite may have carried these molecules until it reached a planet that had all of the conditions conducive to the development and evolution of life, such as the planet Earth. The hypothesis of an extraterrestrial origin for life received further support with the detection of interstellar clouds and the discovery of bacteria that may be of Martian origin.

Proline

The scientific discipline for the study of extraterrestrial life is called exobiology. One day, perhaps, we will have exopaleontology!

CHAPTER IV

THE DEVONIAN, "AGE OF FISHES"

A scene of Devonian animal life at Miguasha : the bony fish *Cheirolepis* (above) swimming through a school of spiny fish *Triazeugacanthus*. Burrowed in the mud is the sea scorpion *Pterygotus.*

The Devonian sets the stage for major events in the evolution of life, some of which had a lasting impact on the Earth's history.

Most of the major evolutionary events identified by paleontologists concerned the evolution of fishes : the extinction, appearance, and diversification of different fish groups, some of which still have living representatives in our world.

This predominance justifies calling the Devonian the "**Age of Fishes**."

Illustration : François Miville-Deschênes

A CRUCIAL GEOLOGICAL PERIOD IN EVOLUTION

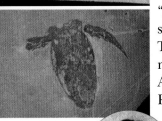

Photo : APM

The Devonian is a geological period that lasted around 54 million years, from 410 to 356 million years ago. The name "Devonian" comes from a county in southwestern England called Devon. There, in the early 19th century, the British geologists Adam Sedgwick and Roderick Murchison identified sedimentary rocks dating from that period in time. Geological formations dating from the Devonian have since been found on all continents.

R. I. Murchison

A. Sedgwick

Sedgwick and Murchison together defined part of the current classification of the Paleozoic's different periods. Their long, fruitful working relationship came to an end because of a dispute, which lasted for years, over the period that would later be named the Ordovician.

Photos : Martin J.S. Rudwick (1985), *The Great Devonian Controversy*, University of Chicago Press.

The Devonian set the stage for major events in the evolution of life, some of which had a lasting impact on the Earth's history. These events had to do as much with the evolution of plants and invertebrates as with the evolution of vertebrates, as well as with ecological and environmental phenomena. There was, for instance, the diversification of land plants, the formation of the first forests, the appearance of terrestrial arthropods, e.g., insects, scorpions, and spiders, the appearance of tetrapods, and the move from water to land by vertebrates.

WHAT IS A FISH?

DEFINITION

According to the Cambridge Encyclopedia, a fish is "any cold-blooded aquatic vertebrate without legs, but typically possessing paired lateral fins as well as median fins. There is a 2-chambered heart, a series of respiratory gills present throughout life in the sides of the pharynx, and a body usually bearing scales and terminating in a caudal (tail) fin."

The words "typically" and "usually" should be highlighted. Indeed, the characteristics of the different fish groups in the Devonian overstep the bounds of this definition. The variants in fish morphology have enabled biologists and paleontologists to establish a phylogenetic tree[1] illustrating the kinship between these groups. The following is a brief description.

Previous page, above: from the sedimentary rocks of the United Kingdom, these Devonian fossil fishes, *Coccosteus* and *Pterichthys*, were among the first to have been described in the 19th century.

On this water-colour of a bowfin (*Amia calva*) are indicated the main external anatomical structures of a fish.

[1] see figure page 85

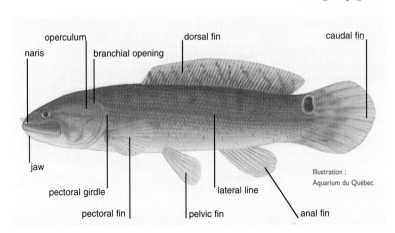

naris

operculum

branchial opening

dorsal fin

caudal fin

jaw

pectoral girdle

pectoral fin

lateral line

pelvic fin

anal fin

Illustration :
Aquarium du Québec.

Complete specimen of *Dinaspidella elizabethae*, a heterostracan agnathan.

Photo : Mark V. H. Wilson, University of Alberta

FROM THE MOST PRIMITIVE TO THE MOST EVOLVED

1 cm

Phylogenetic tree of the major lower vertebrate groups. The most primitive forms are at the top (e.g., the Myxiniforms) and the most evolved ones at the bottom. Only the Myxiniforms, the Arandaspids, and the Petromyzontiforms have no Devonian representatives. Relations of kinship between the groups are indicated, as well as the geological periods of each group (dark bars). The three illustrations depict fossil fishes from Miguasha. From top to bottom: the osteostracan *Escuminaspis laticeps*, the placoderm *Bothriolepis canadensis*, and the actinistian *Miguashaia bureaui*.

[1] see page 58

The diversity of fishes breaks down into major groups that can be classified from the most primitive to the most evolved. A major division is the distinction between the agnathans and the gnathostomes. The agnathans are jawless fishes that appeared in the Ordovician, if not in the Cambrian, over 500 million years ago.[1] Their primitive mouth is no more than a sucking or filter-feeding orifice. There are nine distinct groups of agnathans, only two of which exist today. They are the hagfish, the Myxiniforms, with 43 species, and the lampreys, the Petromyzontiforms, with 40 species. There are very few fossil representatives of either group. The seven other groups of agnathans, the arandaspids, the heterostracans, the anaspids, the galeaspids, the osteostracans, the pituriaspids, and the thelodonts disappeared during the Devonian or before. Most of these groups had a specific geographical range, i.e., their fossils are found only on certain continents.

The galeaspids are known only from China and Vietnam, and the osteostracans from Europe, North America, and part of

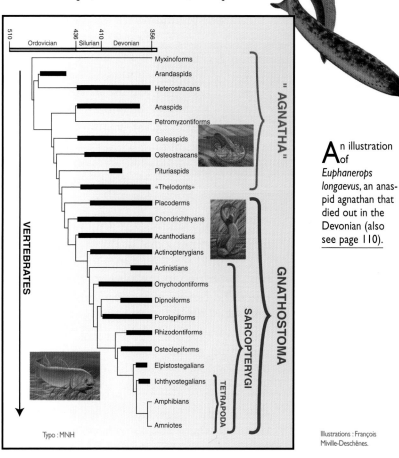

510 Ordovician 436 Silurian 410 Devonian 356

Myxinoforms
Arandaspids
Heterostracans
Anaspids
Petromyzontiforms
Galeaspids
Osteostracans
Pituriaspids
«Thelodonts»

"AGNATHA"

Placoderms
Chondrichthyans
Acanthodians
Actinopterygians
Actinistians
Onychodontiforms
Dipnoiforms
Porolepiforms
Rhizodontiforms
Osteolepiforms
Elpistostegalians
Ichthyostegalians
Amphibians
Amniotes

GNATHOSTOMA

SARCOPTERYGII

TETRAPODA

VERTEBRATES

Typo : MNH

An illustration of *Euphanerops longaevus*, an anaspid agnathan that died out in the Devonian (also see page 110).

Illustrations : François Miville-Deschênes.

Russia. The pituriaspids are rare finds, having been discovered only in Australia. For some years now, following a series of discoveries in Canada's Northwest Territories, it has been suggested that the thelodonts were transitional forms between jawless and jawed fishes.

The gnathostomes, or jawed vertebrates, encompass five major groups. In evolutionary order, they are the placoderms, the chondrichthyans, the acanthodians, the actinopterygians, and the sarcopterygians. Humans belong to the last group.

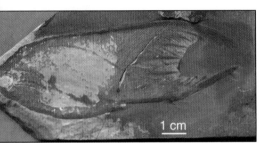

An example of a thelodont: *Sphenonectris turnerae.*

Photo : Mark V. H. Wilson, University of Alberta.

The placoderms, or "armoured fishes," have bony plates covering their head, trunk, and sometimes their pectoral fins. Appearing in the Silurian, they dominated the Devonian's waters until they went extinct at the very beginning of the Carboniferous period. Over 600 species, occupying several kinds of aquatic environment, came into being during their reign of some 75 million years.

Within this group was one of the largest predatory fishes to have lived on our planet, the *Dunkleosteus*, which could reach 7 metres in length.

The chondrichthyans, or "cartilaginous fishes" appeared in the Silurian and are today represented by sharks, rays, and chimera. Their internal skeleton is mainly cartilage, the only fossil bones being sharp fin spines. The chondrichthyan fossil record is especially rich in teeth, which are numerous and made of bony tissue, unlike the rest of the body.

The oldest complete shark skull, around 390 million years old, was recently discovered about forty kilometres from Miguasha, in New Brunswick. The chondrichthyan group is the only one of the gnathostomes not represented at the Miguasha cliffs.

The acanthodians, or "spiny fishes," can be recognized by the small scales that

The Devonian placoderm *Dunkleosteus* from the American Mid-West (above: a cast of its head) could reach 7 metres long.

Photo : APM

Cladoselache, a shark that lived in the Upper Devonian. It was discovered in the Cleveland Shale of Ohio.

Photo : John G. Maisey, American Museum of Natural History

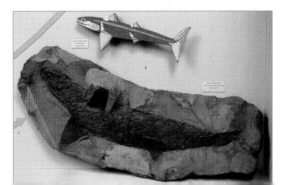

cover their head and body, as well as the presence of sharp spines at the base of all of their fins except the tail. Few deposits yield complete undisturbed acanthodian specimens from the Devonian period. In this regard, Miguasha is remarkable. The group was never really an evolutionary suc-

cess and went extinct in the Permian, around 270 million years ago.

Today, the actinopterygians, or "ray-finned bony fishes," with 29,000 living species, make up nearly 60% of the world's total biodiversity in water and land verte-brate species. In the Devonian, the diversity of this group was much more limited: about fifteen species, most of which are known almost exclusively from isolated fossil scales.

A few representatives of the major tetrapod groups, all present in Quebec. Above: an amphibian, the Eastern grey tree-frog (*Hyla versicolor*). Below: a reptile, the painted turtle (*Chrysemys picta*)…

Photos : Pierre Etcheverry ;
Jacques Trottier

Over 23,000 living species of sarcopterygians, or "lobe-finned bony fishes"

... Left: another reptile, the garter snake (*Thamnophis sirtalis*). Below: a pair of Northern gannets (*Sula bassanus*) during courtship.

are currently known, the overwhelming majority being tetrapods that breathe air and walk on all fours. Nonetheless, there still are a few sarcopterygian fish species, including the dipnoans (lungfish) and the famous coelacanths. The common feature of the sarcopterygians lies in the internal skeleton of their limbs. Whether they have fins or legs, their limbs are joined to the pectoral girdle by a single bone, the humerus— the long bone of the human arm.

The sarcopterygians are divided into four major groups: the onychodonts with 5 species, the actinistians or cœlacanths, with around 125 species, the dipnomorphs, which include the

Photos : Pierre Etcheverry

porolepiforms, with around 25 species, and the dipnoans with around 280 species. Finally, the tetrapodomorph group encompasses vertebrates that acquired or were acquiring quadruped locomotion. Within this group are the rhizodonts, with ten or so species, the osteolepiforms, with around 60 species, the elpistostegalians, with 3 species and, finally, the tetrapods.

The beluga (*Delphinapterus leucas*) is a mammal that can be found in the St. Lawrence waters of Quebec.

Photo : Pierre Etcheverry

The onychodonts, the porolepiforms, and the elpistostegalians died out in the Upper Devonian. The osteolepiforms and the elpistostegalians are considered to be transitional groups between fish and early tetrapods. Miguasha is one of the most representative and diverse fossil sites for this major group of lobe-finned fishes.

CHAPTER V

MIGUASHA,
370 MILLION YEARS AGO

This *Bothriolepis* specimen has seen many geological changes come and go before being discovered, perfectly preserved, in the Miguasha cliffs.

Photo : APM

Quebec's geological history began 2.9 billion years ago, according to dating of the oldest rocks on its territory. Such an age is considerable. The oldest rocks found so far on our planet are 4.4 billion years old, or around 200 million years younger than the earth itself.

Quebec's current landscape has developed through several successive stages.

QUEBEC'S ROCKS TELL US...

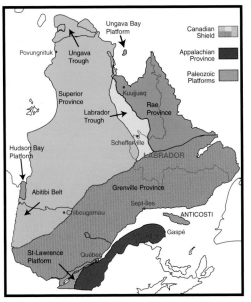

About 95% of Quebec's land surface is a "shield" of mainly igneous and metamorphic Precambrian rocks 2.9 to 1 billion years old, called the Canadian Shield. Although such rocks seldom contain fossils, they may tell us about that era's continental drift and tectonic activity, i.e., the changes in landmasses as a result of moving continental plates. The shield also has a few outcrops of sedimentary rocks containing stromatolites.

This map of Quebec's major geological regions illustrates the long history that has shaped the province's landscape. Miguasha belongs to the Appalachian region, which covers much of southern Quebec.

From P.-A. Bourque's website

[1] cf page 79
[2] cf page 62

Quebec's second major geological episode came with the Paleozoic era. This was around 500 million years after the Precambrian Shield had fallen into place and it saw the formation of the St. Lawrence Platform and the Appalachians. The Precambrian Shield plate collided twice with adjacent European plates, once in the Ordovician and again in the Devonian. The folds and faults visible in cliffs along the northern Gaspé Peninsula

attest to the two big collisions, which occurred
over a period of around
180 million years. These
major upheavals of the
earth's crust pushed up
an immense mountain
chain—the Appalachians
—running from Alabama
in the American south,
through Quebec's
Eastern Townships and Gaspé Peninsula,
and into Newfoundland in the north.

The Gaspé Peninsula bears many marks of different stages of Quebec's geological development. The rock fold, above, is a sign of the deformations caused by the formation of the Appalachians.

The Appalachians have three geological components. The oldest one is composed of Cambrian and Ordovician rocks
and is clearly exposed along the south

shore of the St.
Lawrence from
Quebec City to Cap-
des-Rosiers, on the
Gaspé Peninsula. It
predates the first
collision, so its
rocks are quite
deformed. The
rocks of the Gaspé
Peninsula's interior
and much of the
Eastern Townships
have undergone
less strain, having
formed during the

Sainte-Anne Falls in Gaspé Park. It is made of pillow basalt formed by volcanic lava cooling rapidly in water 500 million years ago.

Photos : APM

Silurian-Devonian between the two collisions. Finally, the youngest component developed after the second collision, in the Carboniferous 330 million years ago, and is hardly deformed at all. It can be seen in the reddish layers that characterize the Percé region as far as Miguasha. For its

The famous Percé Rock (above) at the tip of the Gaspé Peninsula. Its layers of fossil-rich Devonian limestone run almost up and down.

part, the St. Lawrence Platform is composed of sediments from marine deposits and from material that eroded off the then youthful Appalachians between the Ordovician and Devonian periods. The St. Lawrence Platform and the Appalachians are both mainly sedimentary rock and are thus richer in fossils than is the Canadian Shield.

The visit to Gaspé Park continues. Lac aux Américains (right) is a cirque resulting from Pleistocene glacial erosion about a million years ago.

Photos : APM

We know little about Quebec's geological history during the Mesozoic because the glaciers that once covered the province have scoured its

Glacial striations left by the last advance of the glaciers at Lac Blanchet, in the Gaspé Peninsula.

land surface. During the Quaternary, commonly known as the "Ice Age," several glaciations succeeded each other, the last one ending less than 10,000 years ago! The glaciers stripped away many sedimentary layers that had been deposited during the Paleozoic and the Mesozoic. And with them went the information that the fossils in those layers could have provided. This explains why Quebec has no fossils of dinosaurs—those famous reptiles that ruled the world during the Mesozoic.[1] The great Quaternary glaciations thus shaped and carved the landscape we see today.

Mont Xalibu (left) aptly illustrates erosion due to water runoff. From its summit, one can see Lac aux Américains (previous page). Below: a present-day glacier that has shaped a landscape similar to Quebec's. The Athabasca glacier is in Jasper National Park, Alberta.

Photos 1 & 3 : C G C
Photo 2 : APM

[1] see page 73

MIGUASHA, A MICMAC PLACE-NAME

At the far end of Chaleur Bay, a small inlet on the north shore of the Restigouche River estuary is set between steeply rising headlands of red rock. The Micmac Indians called the cliffs megoasag, meaning "where there are red rocks" [megw(e): red; ash: rock or stone]. Today, two small hamlets in the municipality of Nouvelle are called Miguasha and Miguasha-Ouest. The name occurs as early as 1724 on the Thirsisen map as "Pointe de Gouacha." Not until 1847 was it spelled close to the way it is today.

The red cliffs of Miguasha.

Illustrations: Micmac petroglyphs, representing a fish and a caribou, from "Rock Drawings of the Micmac Indians," Marion Robertson, 1973. The Nova Scotia Museum, Halifax.

Photo : APM

Earth and rock hold a special place in creation-of-life stories told by the Micmac—one of the first indigenous peoples to occupy the Gaspé Peninsula, New Brunswick, Nova Scotia, and Prince Edward Island. According to their legends, *Niskam*, the sun, heated a rock covered with morning dew. At midday, the sun was at its strongest and the rock had become very hot, whereupon *Niskam* took and shaped it into the form of an old woman with the help of the great creator *Kisúkwl*. Thus was created *Nukumi*, the original grandmother. The story does not tell us whether the rock came from the Miguasha cliffs.

MIGUASHA TODAY

With its back to the Gaspé highlands and its face to the Restigouche estuary, the Miguasha fossil site forms a narrow plain 40 metres above sea level on average. To the north, the plain runs up against a hill no more than 220 metres high and made of Carboniferous rocks belonging to the Bonaventure Formation—typified by the rocks one finds on Bonaventure Island, near Percé. To the south, the plain abuts the river, either as a cliff rarely over 20 metres high or as an alluvial plain. All of this relief resulted from glacial action during the Quaternary. The plain and the river's cliffs belong to a geological unit that contains the Miguasha fossil site and is known scientifically as the "Escuminac Formation."

A close-up (circle) of conglomerate from the Bonaventure Formation.
A conglomerate is formed from pebbles embedded in a matrix of sand.

Photo : APM

Miguasha Point is an agricultural plain lying on top of the Bonaventure Formation. There are two bays, one being the Escuminac Bay (above left) into which the Restigouche River flows.

Photo : Martin Caissy

THE ESCUMINAC FORMATION

This formation is not very big: only 8 km long and around 1 km wide. Its layers tilt either east or west and sometimes up and down. These deformations suggest that the Escuminac Formation's layers first ran horizontally before being folded during the final formative stages of the Appalachians in the Upper Devonian and again during the last collision with Europe in the Carboniferous.

The Escuminac Formation has a total thickness of around 120 metres and is mainly composed of four types of sedimentary rock: sandstones, silts, shales, and laminites. The

SHALE: Hardened clay mud

SANDSTONE: Sand grains held together by calcareous cement

SILT: Very fine sand

LAMINITES: Succession of shale and silt layers

Close-up of laminites: the shale (dark) and silt (light) layers are each less than a millimetre thick. In the centre, the darker line (inside the box) corresponds to the presence of an acanthodian fossil, seen in cross-section.

Photo : Richard Cloutier

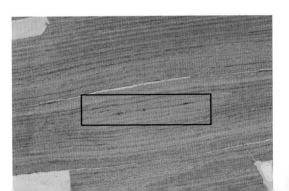

shape of the sand grains in the rock and their variations in size are good indicators of the kind of erosion and transport they have undergone in an aquatic environment. Also occurring locally are layers of conglomerates, e.g., in rivers and along shorelines where the current is relatively strong. All five types of sedimentary rock form layers that vary in thickness. Often, the bottom of a layer will have imprints left by the mechanical action of currents or loose sediments. For example, the current will carry a pebble, a branch, or a piece of bone and leave a furrow in the sand. Or a sudden deposition of sediment may press down into an existing layer and form what is called a load cast. Imprints occasionally affect the thickness of a layer, such as ripple marks left by the action of currents or waves. The Escuminac Formation sediments have many such imprints, thus providing us with invaluable information about past physical environments.

We will now turn to our period of special interest, the Devonian, and to where Miguasha was at that time.

Example of a sedimentary imprint in the Escuminac Formation.

Photos : APM

Ripple marks on a beach today.

Photo : Bruno Vincent, UQAR

[1] See illustration and definition on p. 97

MIGUASHA IN THE UPPER DEVONIAN

SETTING:

A map of the Earth in the Middle Devonian, 390 million years ago.

During the Upper Devonian, the Earth's climate and continents were quite unlike what we know today. The continents were not yet at their current locations. Most were lumped together in the south-

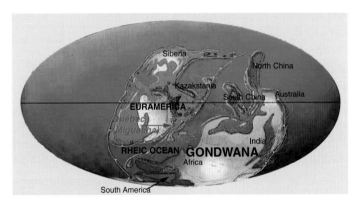

Illustration : C.R. Scotese, 1997, Paleogeographic Atlas, PALEOMAP progress report 90-0497, Department of Geology, University of Texas at Arlington.

ern hemisphere, forming a great paleocontinent called Gondwana, later to give rise to Pangea.[1] Only North America, then linked to Europe and forming the Old Red Sandstone continent, and a few parts of Asia, such as Siberia, Kazakhstan, and China, are thought to have been apart from Gondwana. Quebec lay a few degrees south of the equator, its shores washed by a "paleo-ocean" called the Rheic Ocean. Miguasha was as far from Europe as it now

[1] see page 78

is from Toronto. Its proximity to the equator explains its equatorial and tropical climate during the Devonian—hot and humid.

This general picture has been filled in by very specific analyses of the Miguasha cliffs. With the results of these studies in hand, scientists can describe the environment of this fossil site 370 million years ago in finer detail.

AN ESTUARY 370 MILLION YEARS AGO!

Until the early 1990s, most scientists thought that the sediments of the Escuminac Formation were laid down in a lacustrine environment, i.e., in a freshwater lake. They based their assumption on the presence, underneath the Escuminac Formation, of a special geological assemblage called the Fleurant Formation—a conglomerate of sediments probably carried to their location by river water. A lacustrine origin is also suggested by the absence of fossils of typical marine invertebrates, like corals for example, and typical marine fishes, like chondrichthyans, among specimens found at Miguasha.

The transition is clearly visible here between the darker-coloured Fleurant Formation, at the base, and the Escuminac Formation, above.

Photo : APM

Other scientists, however, have argued for a seacoast environment, as evidenced by fossils of fishes and typical marine organisms in the cliffs, as well as by tidal drift. A new interpretation has recently been proposed to reconcile these diverging lines of evidence.

Kau Bay on Halmahera Island in Indonesia displays a series of characteristics (geology, water chemistry, climate) probably much like those that prevailed at Miguasha in the Upper Devonian. Often considered to be its present-day analogue, this bay helps us better imagine the conditions of life that may have existed in the estuary 370 million years ago.

Photo : Rien Dam

It seems that the Miguasha rocks, fauna, and flora were in an estuary — the mouth of a river on a sea, where freshwater comes into contact with saltwater. Estuary water is brackish, i.e., intermediate in salinity between freshwater and seawater. It is thus conducive to a broad range of aquatic fauna and flora. Estuaries, like oceans, are also subject to the influence of tides. The "intermediate" status of estuaries, with characteristics of both sea and river environments, is consistent with all paleontological and geological data from the Escuminac Formation.

This "paleo-estuary" should not be confused with the present-day Restigouche estuary, where Miguasha is now located.

If we were to go back 370 million years in time, we would see a large freshwater river comparable in size to today's Nile and emptying into the Rheic Ocean through the paleo-estuary. Further inland would be the lofty peaks of the Appalachians, not yet heavily eroded, and offshore, poking through the surface of the Rheic Ocean, the remains of volcanic islands that had been active a few tens of millions of years earlier.

In a world bereft of amphibians, reptiles, birds, and mammals, this was the setting for a major evolutionary advance in the race for the first vertebrate to leave water.

The shores of the Restigouche estuary are now bordered only by a few mounts or hills — the eroded remnants of volcanoes that in the Upper Devonian stood offshore in the Rheic Ocean. Mount Sugarloaf in New Brunswick was a volcano that had already been extinct for several million years in the days of the Miguasha fishes.

Photo : APM

THE FAUNA AND FLORA OF MIGUASHA

Photos : APM

The alternating layers of shale (black and crumbly) and silt (beige and solid) are easily observable on this photo of the Miguasha cliffs.

[1] Marine microorganisms, see p. 106

The Miguasha paleontological site owes its scientific renown to a wealth of fossil fauna and flora from the Devonian. The figures speak for themselves: 7 species of plants; 20 types of palynomorphs (plants known exclusively in a microscopic form); 70 species of spores; 15 genera of acritarchs,[1] 8 species of invertebrates, and 20 species of fish. Such diversity, combined with the quantity of specimens and the quality of their preservation, makes Miguasha unique.

The species preserved in the Escuminac Formation cliffs are given here in evolutionary order with the most primitive forms first. This is their family tree, their tree of life.

THE FLORA

Vegetation diversified in the Devonian, as ancestral horsetails, ferns, and club mosses colonized the coastal continental lowlands. The aquatic environment boasted a variety of one-celled and many-celled algae.

Legends on page 105 :

a) b) Isolated specimens of *Protobarinophyton* - Photos : Patricia Gensel, University of North Carolina at Chapell Hill
c) A frond of *Archaeopteris halliana* - Photo : Patricia Gensel
d) and e) A reconstition of *Archaeopteris halliana* and close-up of its leaves and reproductive bodies containing sporangia and spores (*cf* page 106) - Drawings : Patricia Gensel
f) A present-day fern from Quebec: *Polypodium virgianum* - Photo : Robert Chabot, UQAR

PROTOBARINOPHYTON:
PRIMITIVE PLANTS

Still poorly known because few specimens have been discovered, *Protobarinophyton* was a small primitive plant a few centimetres tall. It likely lived along the shores of bodies of water. Its fossils are in such good condition that we will one day study its reproductive bodies, together with the plant itself.

ARCHAEOPTERIS:
EMBLEM OF THE FIRST FORESTS

The tree fern, *Archaeopteris*, is the most abundant fossil plant in Miguasha sediments. It belongs to a group of plants called the progymnosperms, which paleobotanists consider to be an intermediate link between true ferns on the one hand and seed-bearing conifers and flowering trees on the other.

Archaeopteris had a trunk up to 7 metres tall and a crown of branches bearing compound leaves. Leaf size can be used to distinguish the two Miguasha species from the *Archaeopteris* genus. These abundant trees formed the first forests to be present throughout the world during the Upper Devonian, their remains having been deposited on several Devonian paleocontinents: the Old Red Sandstone continent, Gondwana, Siberia, and China.

SPERMASPORITES:
THE OLDEST SEED-BEARING PLANT

Spermasporites was a small plant, a few centimetres tall, that likely grew along the shores of the estuary. It belonged to a group called the barynophytales, whose Miguasha specimens are still poorly understood. Nonetheless, it is thought that *Spermasporites* may be the oldest seed-bearing plant known to date.

a

MILLIONS OF SPORES!

Spores measure a few thousandths of a millimetre, or microns, across and are the reproductive bodies of primitive plants, i.e., the equivalent of pollen in flowering plants. The millions of fossil spores in the Escuminac Formation belong to over 70 species.

Besides their importance as components of biodiversity, the spores at Miguasha can date specimens from the Escuminac Formation. Spores belong to a special category of "guide" or "index fossils". Such fossils must have a broad geographical range, this being this case with spores because wind and water disperse them over the entire planet. Palynologists specialize in the study of these microfossils and have prepared reference tables that can match a geological period to specific assemblages of fossil index from that period's sediments. Much of the relative dating of rocks from the Devonian to the present is based on these small, barely perceptible fossils.

The Miguasha fossils have thus been dated to the Middle Frasnian—a subdivision of the Upper Devonian.

sporange

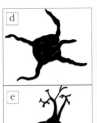
c

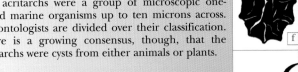

d

f

e

ACRITARCHS: QUESTIONS AND ANSWERS

The acritarchs were a group of microscopic one-celled marine organisms up to ten microns across. Paleontologists are divided over their classification. There is a growing consensus, though, that the acritarchs were cysts from either animals or plants.

These organisms were discovered in Escuminac Formation sediments only in 1996 and their existence is further proof that the Miguasha fishes lived in an environment in direct contact with the sea.

g

a) A spore of *Archaeopteris halliana* enlarged 500 times - Photo : Patricia Gensel

b) A spore-bearing frond of *Archaeopteris obtusa* on which are arranged the sporangia, which contain the spores - Photo : Patricia Gensel

c) Sporangia and spores of a present-day fern: *Polypodium* - Photo : Robert Chabot

d) to g) Four acritarch species, enlarged 500 times: *Dateriocradus*, *Multiplicisphaeridium*, *Muraticavea* and *Veryhachium* - Drawings : Richard Cloutier

INVERTEBRATE FAUNA: IN WATER BUT ALSO ON LAND

Though known chiefly for its many vertebrates, the Escuminac Formation has also yielded invertebrate species. Nearly 70% of current marine invertebrates are soft-bodied organisms that stand little chance of fossilizing. Miguasha invertebrates may have been much more diverse in the Devonian but the full extent of their diversity has not been preserved.

ASMUSIA: BOTTOM OF THE FOOD CHAIN IN THE MIGUASHA ESTUARY

Asmusia is the only fossil crustacean species at Miguasha. It is also the most abundant species in the Escuminac Formation, with shell remains being found from the base to the summit of the formation. Ranging from 2 to 6 millimetres in width, these small crustaceans from the conchostracan group were eaten by many fishes in the paleo-estuary, as attested by the presence of *Asmusia* shells in the digestive tract of numerous fossil fishes, e.g., *Bothriolepis* and *Homalacanthus*, and in frequent coprolites.

1 mm

SEA WORMS: JAWS !

The polychaetes, or sea worms, are still with us today in brackish or fully marine environments. The only trace of them at Miguasha is a jaw fragment, called a scolecodont, around 20 microns across and discovered in a preparation of microorganisms from Escuminac Formation sediments. Sea worm jaws are used in paleontology as indicators of marine influence. This discovery exemplifies the quantity and precision of the work done each day at Miguasha. No specimen, however small, is overlooked!

a) Specimen of a small crustacean: *Asmusia membranacea* - Photo : Thomas Martens

b) *Nereis virens*, a present-day sea worm - Photo : Robert Chabot, UQAR

c) Scolecodont discovered in the Escuminac Formation. Drawing shows it enlarged **500 times** - drawing : Richard Cloutier

EURYPTERIDS: PREY OR PREDATORS ?

The eurypterids were great sea scorpions up to three metres in length and often associated with fish throughout the Devonian. It has been suggested that some eurypterids could make brief forays onto dry land. To date, only fragments of them have been discovered in Escuminac Formation sediments, these being paddles used for swimming and pieces of carapace.

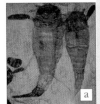

Some paleontologists argue that these great sea scorpions could use their claws to attack the slowest fishes; however, it is also known that they moulted periodically like present-day arthropods, e.g., lobsters, crabs, and scorpions, thereby losing their protective shells and becoming easy prey. So were these great sea scorpions prey or predators of Miguasha fishes? The question remains unanswered.

AMONG THE FIRST LAND SCORPIONS

Unlike the eurypterids, the scorpion *Petaloscorpio bureaui* did not live in water but on dry land, on the estuary's shores. It was one of the oldest known land scorpions and had a size of up to thirty centimetres. Its external anatomy was like that of present-day scorpions and has been extensively studied, thanks to the excellent preservation of many fossil specimens from Escuminac Formation sediments.

Was it venomous? It is hard to say at this time. In addition to *Petaloscorpio*, Miguasha has yielded fragments from another group of scorpions: the Gigantoscorpionidae. As their name indicates, they reached impressive sizes—up to a metre in length.

a) Fossils of *Eurypterus remipes*, a eurypterid of the Upper Silurian from New York state - Photo : Armand Dubé

b) Reconstitution of *Pterygotus* - Illustration : J. Gardner

c) Complete specimen of the scorpion *Petaloscorpio bureaui* - Photo : APM

d) *Petaloscorpio bureaui* drawing of fossil c) - Drawing : Andrew J. Jeram, Ulster Museum, Belfast, Ireland

e) A present-day scorpion from Queensland - Photo : Pierre Etcheverry

VERTEBRATE FAUNA

Twenty species belonging to eleven major groups of lower vertebrates have been discovered in the Miguasha cliffs. Though living in the same environment over the same period of time, these species attest to different stages of vertebrate evolution. Most belonged to a large number of groups that died out many millions of years ago. The others survive among the most primitive forms of present-day groups. Such fauna are a real mine of information for understanding the evolution of fishes and the origin of tetrapods—our own group even though we walk on two legs.

Legends of the three next pages:

Page 110:
a) A lamprey, whose mouth can be clearly seen - Photo : Pierre Etcheverry
b) Complete specimen of the anaspid *Euphanerops longaevus* and a drawing of this fish - Illustration : Richard Cloutier

Page 111:
c) Complete specimen of the osteostracan *Escuminaspis laticeps* - Photo : APM
d) Reconstitution of *Escuminaspis laticeps* - Illustration : François Miville-Deschênes

Page 112:
e) Complete specimen of the placoderm *Bothriolepis canadensis* - Photo : APM
f) Reconstitution of two *Bothriolepis canadensis* - Illustration : François Miville-Deschênes

ANASPIDS: ANCESTRAL LAMPREYS ?

Specimens from two species of anaspids, ***Endeiolepis aneri*** and ***Euphanerops longaevus***, have been discovered at Miguasha. Slim and spindle-shaped, the anaspids did not have a rigid bony shield covering their head, unlike other groups of agnathans or jawless fishes, but rather long, fine scales. They swam by using a downward-pointing tail and a pair of long ventral fins.

Only twenty or so anaspid species are known from the 70 million years starting with their origin in the Lower Silurian and ending in their extinction in the Upper Devonian. *Endeiolepis* and *Euphanerops* are the last known representatives. We should add that, of all agnathans, the anaspids have probably stirred up the most controversy over their anatomy, behaviour, and evolutionary status. Some paleontologists consider the forms discovered at Miguasha to be related to present-day petromyzontiforms, which lampreys belong to (see page 85). Indeed, *Euphanerops* looks like a lamprey in its mouth structure—a circle of cartilage lined with minuscule denticles.

OSTEOSTRACANS:
ON THE EVE OF THEIR EXTINCTION

The osteostracans, another group of jawless fishes, had a horseshoe-shaped head protected by a shield of small bony plates, a pair of paddle-shaped pectoral fins, and a body covered with either bony plates or tiny scales invisible to the naked eye—like the Miguasha specimens. The eyes were on top of their head, with the mouth and gill opening beneath. Such characteristics indicate a benthic mode of life, i.e., they lived on the water bottom, moving about and feeding on the mud.

The Escuminac Formation sediments contain two kinds of osteostracans. *Escuminaspis laticeps* could reach a length of 80 centimetres whereas *Levesquaspis patteni* was smaller. Few specimens have been discovered because the fragile bones of their skull shield make preservation difficult.

From their origin in the Silurian to their extinction in the Upper Devonian, the osteostracans always lived in waters associated with the Old Red Sandstone continent, on which Miguasha was located. This geographical range explains why no osteostracan fossil has yet been found in the southern hemisphere. The ones at Miguasha are the last known representatives of their group, the cause of their extinction being still unknown.

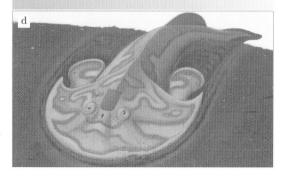

e

PLACODERMS:
KINGS AND RULERS OF THE DEVONIAN

The placoderms, or armoured fishes, were characterized by bony plate armour on their head and thorax. Miguasha has yielded specimens belonging to two placoderm species: *Bothriolepis canadensis* and *Plourdosteus canadensis*.

The nostrils and eyes of *Bothriolepis* were grouped into a single small opening on top of its head. This suggests that **Bothriolepis canadensis** lived on the paleo-estuary bottom, just like the osteostracans described above. Bony plates covered not only the head and thorax but also the pectoral fins, which in turn had bony projections.

Back in 1842, this bony shield fooled the site's discoverer, Abraham Gesner, into thinking he had found a fossil turtle at Miguasha. No scales or plates covered the rear end, so we seldom find fossils that include the tail. Despite these constraints on fossilization, several dozen complete specimens show the elongated appearance of the creature's trunk and tail. Of the thousands of *Bothriolepis* fossils discovered, one has recently revealed an anatomical feature previously unknown in Devonian fishes: the outline of a few blood vessels (see page 46).

f

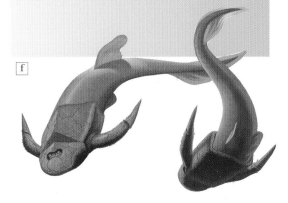

PLACODERMS:
KINGS AND RULERS OF THE DEVONIAN
(continuation...)

Plourdosteus canadensis was a carnivore and probably a voracious predator. The Swedish paleontologist Tor Ørvig named it after the Plourde—a famous family of Miguasha collectors. On the basis of his description, species closely related to the Canadian form have been identified in Scotland, Germany, Latvia, Russia, and Iran.

Such a paleogeographical range is understandable because of the proximity of these different sites to each other during the Devonian, all being part of the Old Red Sandstone continent and bordering on the Rheic Ocean.

Above:
a) Specimen of the placoderm *Plourdosteus canadensis* - Photo : APM
b) Reconstitution of *Plourdosteus canadensis* - Illustration : François Miville-Deschênes

Next page:
c) Reconstitution of the acanthodian *Triazeugacanthus affinis* - Illustration : François Miville-Deschênes
d) Reconstitution of the acanthodian *Homalacanthus concinnus* - Illustration : François Miville-Deschênes
e) Specimen of the acanthodian *Homalacanthus concinnus* - Photo : APM
f) Specimen of the acanthodian *Diplacanthus horridus* - Photo : APM

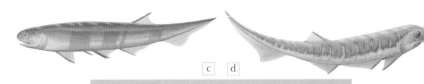

c   d

e

ACANTHODIANS: A THORNY PROBLEM

Of the major groups of fishes, only the chondrichthyans—today represented by sharks, rays, and chimeras—are not represented in the Miguasha sediments. Like today's cartilaginous fishes, most Devonian sharks lived in a marine environment and should not have ventured into the estuary waters of the Escuminac Formation. The Miguasha fauna nonetheless include a group of fishes, the acanthodians, that are sometimes nicknamed "spiny sharks" because they have some morphological resemblance with sharks. Unlike true sharks, the acanthodians had a bony skeleton.

Acanthodians are easily distinguished from all other fishes by the sharp projections at the base of each fin, except the tail. Despite these bony spines, they have nothing in common with the present-day sticklebacks of Quebec freshwater lakes and streams.

Four species are represented among the vertebrates of the Escuminac Formation: *Diplacanthus ellsi*, *Diplacanthus horridus*, *Homalacanthus concinnus*, and *Triazeugacanthus affinis*. Their group is the most abundant one in the formation: certain strata of the Escuminac Formation harbour over 600 fossil specimens per square metre. They were small fish from 3 to 50 centimetres in size.

f

ACTINOPTERYGIANS: *CHEIROLEPIS* ON THE THRESHOLD OF EVOLUTIONARY SUCCESS

The actinopterygians, or ray-finned bony fishes, are today the most abundant group of vertebrates on our planet, the number of species approaching 29,000. Ever since their appearance in the Upper Silurian, they have gradually adapted to all available aquatic habitats over a very wide range of temperatures, salinities, and water depths. They are now highly diverse and still evolving.

There were just fifteen or so species in the Devonian. One was ***Cheirolepis canadensis***, the only actinopterygian from the Escuminac Formation. It is considered to be one of the most primitive species of the group and also one of the best known. Despite a limited number of specimens, the anatomy of *Cheirolepis* has been relatively well studied. Of all fossil species at Miguasha, it is the one that most closely resembles present-day fishes in its shape—much like that of some freshwater minnows and mullets in Quebec today

Above:

a) Specimen of the actinopterygian *Cheirolepis canadensis* - Photo : APM

b) Reconstitution of *Cheirolepis canadensis* - Illustration : François Miville-Deschênes

Next page:

c) Juvenile specimen of the actinistian *Miguashaia bureaui* - Photo : APM

d) Reconstitution of the actinistian *Miguashaia bureaui* - Illustration : François Miville-Deschênes

SARCOPTERYGIANS: MOVING ONTO DRY LAND

The sarcopterygians, or lobe-finned bony fishes, may be said to encompass as many species as belong to the actinopterygians. Most of this diversity would cover thousands of species of tetrapodomorphs, some 24,000 of which are tetrapods that today live on dry land—including our own species. Three of the four major sarcopterygian groups are present in the Miguasha fossil record.

Of all known fish groups, only the sarcopterygians are like tetrapods in having a single bone at the base of the pectoral and pelvic fins. This single bone is the humerus for the fore-limbs and the femur for the hindlimbs. The evolution of these fin bones was a major step in the move to life on dry land.

MIGUASHAIA:
THE MOST PRIMITIVE CŒLACANTH

Miguashaia bureaui, the only actinistian from the Escuminac Formation, was among the first cœlacanths to appear in the course of evolution. It was 60 centimetres long, making it the largest cœlacanth species of the Paleozoic, before actinistians up to 4 metres and more in length appeared in the Mesozoic. Miguashaia bureaui is one of the few fossil fishes whose morphology is fully known, both as a juvenile and as an adult.

A few fragmentary remains of another type of *Miguashaia* have been found in Latvia and a primitive cœlacanth species, Gavinia, has been reported from Australia. Despite these recent discoveries, the Miguasha cœlacanth currently tells us the most about the origins of this fish group, which held so many surprises for us in the past century.

LATIMERIA
AND THE QUEST FOR LIVING FOSSILS

One of the most incredible sagas in the history of biology has certainly been the discovery of the cœlacanth.

*L*atimeria chalumnae (below) and cast of this coelacanth (page bottom).

Paleontologists had known cœla-canths since the 1840s but only as fossils. In 1938, a fish-ing boat off the South African coast hauled up a strange-looking fish in its nets. Intrigued, the curator at the Museum of East London, in South Africa, sent drawings of the fish to J.L.B. Smith, an ichthyologist in Cape Town. Smith was stupefied, realizing that the caught fish was the only living form of a cœlacanth—assumed to have gone extinct when the dinosaurs did 65 million years ago. Not until 1952 was a second one discovered off the Comoros Islands, where most subsequent specimens have been caught. It was named Latimeria chalumnae.

Photo : Eugene Balon, Guelph University,

Studies of *Latimeria* have shown that this living fossil has evolved little since the beginning of the Devonian, when the first cœlacanths appeared.

The discovery of a living fossil is a rare event in itself. It is even rarer for it to happen twice within the same group of organisms. Yet that is what happened recently. A new coelacanth species *Latimeria menadoensis* was discovered in July 1998, this time in Indonesia, north of the island of Sulawesi and over 10,000 km from the Comoros Islands.

Photo : APM

LUNGFISH:
FROM GILLS TO LUNGS

Scaumenacia curta and *Fleurantia denticulata* are two species of dipnoi, or lungfish, from Miguasha.

They are among the best known lungfish species, thanks to the quality of some fossils from the Escuminac Formation. Several hundred specimens of Scaumenacia curta are preserved in three dimensions, revealing all of the fine skeletal details of the body and fins.

Most lungfish, including *Scaumenacia*, have highly unusual dentition. Pairs of dental plates cover part of the roof of their mouth and their lower jaw. Each of the four tooth plates is formed from successive rows of increasingly bigger teeth that appear as the animal grows. Most fishes lose their teeth in the course of life but lungfish keep all of theirs, which are worn down through use and become plates.

Like present-day lungfish, *Scaumenacia* had to grind its food with these interlocking slabs.

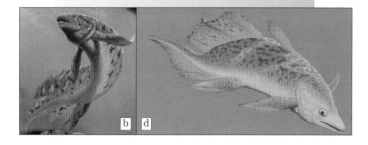

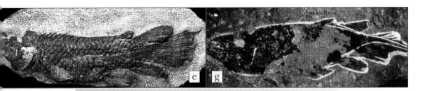

POROLEPIFORMS: POOR RELATIONS OF THE SARCOPTERYGIANS

The porolepiforms were a not very diverse group of sarcopterygians related to the lungfish. They went extinct in the Upper Devonian. Three porolepiform species have been discovered in the Miguasha cliffs: *Holoptychius jarviki*; *Quebecius quebecensis*, which could reach a metre in length; and a third species not yet described.

As with lungfish, the pectoral fins of *Holoptychius* were highly lobed, a characteristic that has earned the sarcopterygian group the name of lobe-finned fishes.

Page 118:
a) Specimen of the lungfish *Scaumenacia curta* - Photo : APM
b) Reconstitution of *Scaumenacia curta* - Illustration : François Miville-Deschênes
c) Skull of the lungfish *Fleurantia denticulata* - Photo : APM
d) Reconstitution of *Fleurantia denticulata* - Illustration : François Miville-Deschênes
   Above:
e) Specimen of the porolepiform *Holoptychius jarviki* - Photo : APM
f) Reconstitution of *Holoptychius jarviki* - Illustration : François Miville-Deschênes
g) Specimen of the porolepiform *Quebecius quebecensis* - Photo : APM
h) Reconstitution of *Quebecius quebecensis* - Illustration : François Miville-Deschênes

a

EUSTHENOPTERON FOORDI: A MAJOR EVOLUTIONARY LINK

Two groups of sarcopterygians now extinct, the osteolepiforms and the elpistostegalians, have shed much light on vertebrates and their transition from water to land. The osteolepiforms are represented in the Escuminac Formation by **Eusthenopteron foordi** and **Callistiopterus clappi**, the latter being known from a single small specimen preserved at the Harvard University Museum of Comparative Zoology.

First discovered in 1879, *Eusthenopteron foordi* soon caught the attention of scientists because of the anatomical similarities between this large fish, sometimes over a metre in length, and the primitive tetrapods that began to appear in the Upper Devonian.

The Canadian paleontologist Whiteaves noted that its pectoral fins had a radius and a cubitus, in addition to a humerus at the base. Both bones formed the second fin segment, corresponding to the human forearm. As well, the humerus had a series of small bony processes, thus allowing better insertion of the muscles and making the fins more manoeuvrable. In other words, these fins were the forerunners of true feet. Such a remarkable feature inspired the name that Whiteaves gave to the fish, "eusthe" meaning robust in Greek and "pteron" meaning fin.

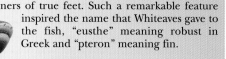

b

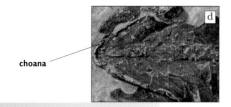

choana

The Swedish paleontologist Erik Jarvik pointed out the presence of internal nostrils, or choana, in *Eusthenopteron*. This opening was at the front of the roof of the animal's mouth and was connected by a passage to the external nostril, thus enabling *Eusthenopteron* to breathe air from the atmosphere while keeping its mouth closed. Such a change was an evolutionary breakthrough in developing respiration that could cope with life out of water. The fin and nostril bones are but two examples from a series of characters that early tetrapods and *Eusthenopteron* had in common.

They shared similarities at the microscopic level, too, such as the folded dentine structure of their teeth. Despite lungs for respiration and sturdy fins, the overall anatomy of *Eusthenopteron* did not allow it to walk or even crawl out of water.

Considered for a century to be an intermediate link between water and land vertebrates, *Eusthenopteron* still excites interest because of its exceptional characteristics and it ranks among the most thoroughly studied of all fish species. It has contributed greatly to shaping modern views on the evolution from fish to land vertebrates.

Pages 120, 121:

a) Specimen of the sarcopterygian *Eusthenopteron foordi* - Photo : APM

b) Reconstitution of *Eusthenopteron foordi* - Illustration : François Miville-Deschênes

c) Skull of *Eusthenopteron foordi* preserved in three dimensions - Photo : R. L. Carroll, McGill University

d) Roof of mouth of *Eusthenopteron foordi* with choana, indicated by pointer - Photo: APM
   Page 122: e) bottom and f) top views of a skull of *Elpistostege watsoni* - Photos : APM

g) Complete specimen of *Panderichthys,* found in Latvia, a species closely related to *Elpistostege* - Photo : Ervin Luksevics, Latvian Museum of Natural History, Riga.

h) A crocodile's shape and head are like that of *Elpistostege*: its eyes and nostrils are out of the water while the rest of its body remains submerged - Photo : P. Etcheverry

ELPISTOSTEGE: FISH OR TETRAPOD?

A second group of fossil sarcopterygians has been crucial to our understanding of tetrapod origins: the elpistostegalians. They are considered to be the sister group of the tetrapods in that they and tetrapods share characteristics that are unique to both. The elpistostegalians reached a more advanced evolutionary stage than did the osteolepiforms.

The sole representative of this group at Miguasha is **Elpistostege watsoni**, a large Devonian fish that has to date left only three incomplete specimens. Despite the fragmentary state of its fossils, its affinity with primitive tetrapods is incontestable. *Elpistostege* and *Panderichthys*—an elpistostegalian found in Devonian rocks from Latvia—are the only known fossil or living fishes with a pair of frontal bones like tetrapods. The skulls of both species strikingly resemble those of primitive Devonian tetrapods, despite their having gular plates—bones found only in the lower jaws of fish.

Like early tetrapods, *Elpistostege* had eye sockets fairly close to each other on top of its head. Its external nostrils were near the edge of its mouth and its skull was relatively flat. It could thus raise its eyes and nostrils above water, thereby making vision and breathing out of water possible, while keeping the rest of its body submerged.

We have no direct information on *Elpistostege*'s body and fins, so we must deduce its overall appearance from *Panderichthys* morphology. Such as *Eusthenopteron*, *Panderichthys*—and hence *Elpistostege* by inference—lacks certain anatomical features that would have enabled it to walk. Though aquatic, *Elpistostege* must have lived at the surface of the Miguasha paleo-estuary, periodically raising its head out of water. In this species and others, we can see vertebrates slowly moving away from an aquatic environment.

THE MIGUASHA FOSSILS AND THE ORIGIN OF TETRAPODS: A KEY TO THE PUZZLE

The sarcopterygians of Miguasha, especially *Eusthenopteron* and *Elpistostege*, prove that the evolution from fishes to tetrapods did not occur in one sudden leap but rather gradually over several million years. About a decade ago, paleontologists were still thinking that Devonian tetrapods—the descendants of sarcopterygian fishes—had adapted to life on land from the beginning. Since the early 1990s, this assumption has been challenged by the discovery of close to ten new species of Devonian tetrapods around the world, almost as old as the Miguasha fossils, thus shedding new light on this evolutionary stage. These early tetrapods had feet with as many as eight toes, a strong vertebral column, a rib cage, and no dorsal or anal fins. They looked like large salamanders up to a little over a metre in length and had retained some fish-like characteristics, such as well-developed gills for breathing under water and a tail for swimming.

With little information being yet available on *Elpistostege*, this reconstitution is based mainly on data from *Panderichthys*, a closely related species discovered in Latvia. *Elpistostege* probably lived along the estuary's shores.

Bottom of the page: forelimb of the tetrapod *Acanthostega gunnari*, which lived 360 million years ago. Note the presence of eight toes.

Illustration : François Miville Deschênes.

Photo : Jennifer A. Clack, University Museum of Zoology, Cambridge
Illustration : Richard Cloutier.

Reconstitution of *Ichthyostega*, an Upper Devonian tetrapod from east Greenland

Photo : APM

By studying the foot bones of two Late Devonian primitive tetrapods from Greenland, *Acanthostega* and *Ichthyostega*, the British paleontologists Jennifer Clack and Michael Coates have clearly shown that these early quadrupeds used their feet not for walking on dry land but rather for swimming in the dense vegetation of shoreline wetlands, swamps, and deltas.

Tetrapods, therefore, would have used their limbs for moving about on the ground much later, the true move to dry land taking place over 20 million years after they first appeared.

So the first tetrapods were just funny-looking fish, perhaps much like the ones that lived at Miguasha 370 million years ago.

Hind foot of the tetrapod *Tulerpeton*, which lived 360 million years ago and was found in Russia.

Photo : Museum national d'histoire naturelle de Paris

CHAPTER VI

WHAT THE MIGUASHA FOSSILS TELL US

When the work begins and where it really matters: digging at the Miguasha cliffs.

A fundamental concept in paleontology, as in geology, is uniformitarianism, i.e., the present is the key to the past. It was put forward over 100 years ago and is based on a simple principle: the biological mechanisms in effect today are the same as the ones that governed living things in the past.

Thus, to understand the life of fishes that died out 370 million years ago, paleontologists study analogous fishes from modern species. Through this approach, we can see the Miguasha fossil specimens as organisms that hatched, grew, struggled to survive, and died.

HOW THE FISHES GREW

To study the growth of living animals, one need only observe them for several weeks and note the changes that affect their size and morphology, hour by hour, day by day. Of course, today's fossils were once living organisms that grew and changed with age. Studying their growth, however, is not so simple.

Example of a growth series used for studying the placoderm *Bothriolepis*.

The approach used by paleontologists is comparable to the one used by biologists on living animals. It involves measuring the increase in an animal's size—the clearest sign of its growth. A paleontologist, however, will measure a series of different-sized fossils from the same species.

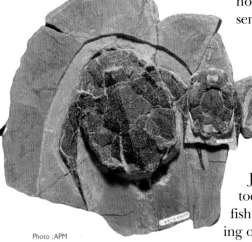

Photo : APM

Just like fish today, prehistoric fish were continually growing organisms, i.e., they grew all life long, with their rate of growth falling off after sexual maturity. So, within any given species the size of a fish will be proportional to its age. One can deduce, then, that the smallest specimens in a series were the youngest, perhaps even larvae, and that the largest specimens were the oldest at the time of death. Again, a series of different-sized fossils is needed to make the approach work. Few sites with fossils from the Devonian, or from any geological period, provide adequate material for study. In this sense as in others, the Miguasha site is exceptional: the Escuminac Formation has yielded very small and large specimens for ten of the twenty species of fish present. For example, the fossils of the placoderm *Bothriolepis canadensis* range from seven millimetres to over forty centimetres long. The quantity of these fossils, and their quality of preservation, has enabled us to understand and describe some of the changes that extinct species went through during growth.

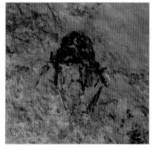

Photo : Robert Chabot

The smallest known specimen of *Bothriolepis*, whose bony shield measured 7 mm across.

Among the Miguasha fish, *Eusthenopteron foordi* has drawn the most attention from paleontologists looking into growth. In the late 1960s, K. S. Thomson and K. V. Hahn studied tail shape changes in around ten specimens of *Eusthenopteron*. They noticed that the central portion of the tail was longer in the smallest individuals, forming a sort of caudal whip, and shrank relatively early in life, thus causing

Hans-Peter Schultze has studied the Miguasha fauna since the early 1970s and is internationally recognized as a specialist on the Escuminac Formation sarcopterygians. His work also takes in the acanthodians, and he has studied and described many species discovered at the Miguasha cliffs, such as *Miguashaia*.

the tail shape to change. This must have probably been accompanied by changes in *Eusthenopteron*'s behaviour, particularly in its mode of locomotion.

In 1984, the German paleontologist Hans-Peter Schultze took a look at *Eusthenopteron*, this time to study the changing proportions of its different skull bones. He noted that several parts of *Eusthenopteron*'s head changed proportionately during growth—its allometry changing with increasing size. The human head, and its changing size, provides us with a contemporary example of allometry. A baby's head measures about a third the size of its body; an adult's only a seventh. "Mature" *Eusthenopteron* have, in proportion to their head size, smaller eyes than do juveniles, whose eyes are enormous. H.P. Schultze thus discovered that *Eusthenopteron* went through transformations that are today common in most fish, i.e., larvae and juvenile fish have big eyes to see their predators better and escape from them.

Photo : Waltraud Harre

A recent inventory of paleontological collections around the world has estimated the number of *Eusthenopteron* specimens at 2,600, their size ranging from around 3 centimetres to a bit over a metre in length. The sheer magnitude of this collection encouraged a team of Canadian paleontologists to study a different aspect of the growth of this fossil fish. This recent research work, currently in progress, looks into the order of appearance of bones in the body

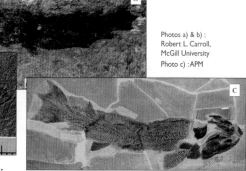

Photos a) & b) :
Robert L. Carroll,
McGill University
Photo c) : APM

of these fish. As they grow, the bones of the vertebral column do not all develop at the same time. They follow very specific "ossification plans," which are in part genetically controlled and in part dependent on the animal's environment. For example, in many larvae of living fish species, the rays of the fins ossify a few days after hatching, later followed by some mouth bones. The order of ossification is genetically controlled, but geared to the larvae's basic needs: movement for escape, and thus survival, and for feeding. Thus, the order of ossification of over 200 bones of *Eusthenopteron* has been established by

Three specimens of *Eusthenopteron*, measuring respectively 3 cm, 5 cm, and 80 cm. By studying them, we can better understand the growth of this Devonian fish.

An example of an ossification plan. Drawing (a) depicts the skeleton of the

comparing a series of specimens of different sizes.

The initial results show that the fin's rays and most of the bones covering the head are

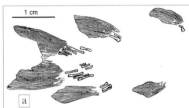

1 cm

a

tail of a young *Eusthenopteron*, with few ossified parts. Drawing (b) shows the skeleton of the tail of an older specimen .

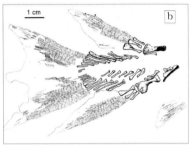

1 cm

b

already present in juveniles 3 centimetres long. Next comes the gradual development of the bones forming the internal tail

d

c

skeleton and then the bones of the other fins. This suggests that, very early in life, *Eusthenopteron* must have already been an active fish, probably a predator. The ossification sequence of young *Eusthenopteron* appears, then, to be fully consistent with a mode of life adapted to their estuary environment and requiring active locomotion.

An equivalent study of present-day fish. Photos (c) and (d) show the fry of Arctic char at, respectively, 26 and 90 days. Blue: cartilaginous parts. Red: ossified parts, which are more numerous at 90 days.

Illustrations a & b : Richard Cloutier & Robert Carroll
Photos: Simon Lamarre, UQAR

HOW THE FISHES FED

In waters around the entire world, over 29,000 fish species occupy practically all possible levels of the food chain. But what about the fishes that lived some 370 million years ago? What can be said about their feeding and behaviour? It is of course impossible to see them in action, but we can get precious clues by studying certain fossils. At this point, the paleontologist has to do some detective work!

Careful examination of fossils provides much information. The shape of the body, the teeth, and the jaw tells us what the fish fed on. The same hints can be gleaned from coprolites and regurgitates, which may yield crustacean or fish fragments, mainly scales. These two fossil categories tell us what a fish ate. For example, a recent study of coprolites from the Escuminac Formation has shown that over a third of them contained acanthodian scales. These coprolites, though, provide no information identifying the fishes that fed on the acanthodians. Such invaluable information can be obtained when a fossil fish is discovered

Dental plates of the lungfish *Scaumenacia curta*, with which it could grind the shells of the small crustaceans it fed on.

Photo : APM

Backdrop to the text: a reconstitution of Miguasha fauna in the Devonian.[1] On this page are shown a *Scaumenacia* in the upper left, four *Cheirolepis* in the foreground, and a *Bothriolepis* in the lower right...

and its digestive system or mouth still has remains of its last meal. Alas, such fossils are rare.

The following description is the result of patient, meticulous work on all of these types of fossils. It helps us imagine what was the life of the species that inhabited the Miguasha paleo-estuary.

The osteostracans *Escuminapsis* and *Levesquaspis* lived at the estuary bottom together with the placoderm *Bothriolepis*, where all of these fishes had to share a common food resource made up of minuscule benthic life-forms: *Asmusia* for example and other microorganisms.

Several *Scaumenacia* specimens have been discovered with a large quantity of *Asmusia* in their digestive tract, proving that *Asmusia* were a choice food item for this dipnoan. It had dental plates with which it could partially grind the small *Asmusia* shells.

Other species fond of *Asmusia* were the small "spiny sharks," such as *Homalacanthus*, which filtered organic matter from the water just as today's basking sharks filter-feed on plankton. These acanthodians were in turn easy prey for predators inhabiting the estuary. A *Homalacanthus* measuring 18 centimetres

[1] see on page 80.

has, for example, been found inside a *Eusthenopteron*, itself nearly 50 centimetres long. Just as a pike will today capture sticklebacks headfirst to keep their spines from puncturing its digestive system, this *Eusthenopteron* had swallowed its acanthodian headfirst before dying shortly after.

Other than *Eusthenopteron*, the most fearsome predators at Miguasha were certainly the placoderm *Plourdosteus*, the actinopterygian *Cheirolepis* and the sarcopterygians *Holoptychius* and *Quebecius*. Although we still have no concrete proof, the morphology of the body and head of the porolepiforms and of *Plourdosteus* suggest that these three major kinds of fish had a predatory lifestyle.

In the case of *Plourdosteus canadensis*, this hypothesis is borne out in particular by the existence of especially efficient tools for cutting up prey—shear-like jaws that were equipped, not with teeth, but with a series of sharp cutting blades.

Its hydrodynamic shape, its excellent mobility, and its jaws armed with numerous teeth mark out the actinopterygian *Cheirolepis* as an effective and fearsome predator, even able to attack fish of its own species. Indeed, scientists have recently linked *Cheirolepis* to one of the oldest known examples of vertebrate cannibalism

... On this page may be seen a *Eusthenopteron* in the middle, and to the right an *Escuminaspis* and an *Elpistostege*.

Illustration: Robert J. Barber (American Museum of Natural History, New York.)

with the discovery of a young fish 17 centimetres long in the mouth of a specimen that must have measured nearly 40 centimetres. *Cheirolepis* was also the prey of other predators in the estuary, in particular *Eusthenopteron*. Remains of *Cheirolepis* have already been found in the stomach contents of some *Eusthenopteron* specimens.

It seems, then, that Devonian life in the Miguasha paleo-estuary was every bit as awesome and terrifying as Jurassic Park!

Two *Plourdosteus*, undoubtedly fearsome predators, here amid a school of small acanthodians.

Illustration: François Miville-Deschênes

HOW THE FISHES DIED

In this reconstitution of the past, paleontologists try to discover not only how the Miguasha fishes lived, but also how they died. Just like today, many different events could cause death. Some light is shed by taphonomy: the study of the processes of fossilization, including the causes of death of organisms and all of the transformations involved in the geological history of a fossil.

Miguasha fossil fish are found in several strata of the Escuminac Formation, sometimes in the form of isolated bones, but very often in the form of complete fish. The body's shape helps paleontologists identify the cause of death. Often, fossils of *Eusthenopteron* and acanthodians show a pronounced curvature along the back. This observation, still valid for fish living today, indicates death from a lack of calcium or a respiratory problem. In fish, these deficiencies lead to bouts of muscular contraction and give the body a curved outline. The high temperature of the Devonian's equatorial climate and the natural turbidity that must have prevailed in the Miguasha paleo-estuary may have reduced the water's oxygen content, thus causing many fish to die.

When a complete fish is discovered in a cliff stratum, excavations are done to see whether other fossils have been preserved at the same level. If so, lengthy research work will begin, since fish from the same layer may have all died simultaneously from a single cause. In certain strata of the Escuminac Formation, up to 600 specimens of *Triazeugacanthus* are present per square metre, indicating successive mass die-offs of this small acanthodian.

Each small dark line on this laminite slab corresponds to the degraded, decomposed remains of a specimen of the small acanthodian *Triazeugacanthus*. This rock layer contains many thousands of them—evidence of a mass die-off of these acanthodians in the Miguasha estuary 370 million years ago.

Photo : Richard Cloutier

Often, the fossils display signs of advanced decay. Hundreds of dead fish found beside each other—we have here the makings of a detective investigation! To understand these Devonian die-offs better, we look for comparable examples among current phenomena. Comparable large-scale mortality arises from natural phenomena, most often seasonal, such as a

Present-day example of a mass die-off. This pond in the northern United States dried up over the winter, thus causing the deaths of thousands of turtles and fish, such as this longnose gar—whose bones can be seen in the foreground. Such mass mortality thus has a natural cause and reoccurs cyclically.

rapid change in the water's salinity or temperature, a drop in the water's oxygen content, or drought. Mass die-offs are of great interest to paleontologists because they shed light on past ecosystems, the paleoenvironment, and possible relationships between organisms and their milieu.

FOSSILS AND HUMANS

The passion for fossils handed down from father to son. Antoine and Euclide Plourde beside the Miguasha cliffs in 1937.

Since 1842, the Miguasha fossil cliffs have been bringing together generations of local people and outsiders, whether scientists, fossil-hunters, or collectors.

Over the years, science and the thirst for knowledge have joined forces in a saga of scientific inquiry—through excavations, discoveries, descriptions, and publications—into Miguasha's fossil fauna and flora and into its paleoenvironment.

B esides the quality of its preservation, this specimen of a young *Eusthenopteron* has a special value. It is one of the first fossils to have been collected at the site. Foord discovered it in 1881, as attested by the handwritten label glued to the rock.

With the support of these men and women, their research, and their writings, the Miguasha heritage site has become world-renowned and has shed new light on the evolution of life on our planet. In addition, the Miguasha cliffs have been extremely generous towards those who have come to it for answers. It has offered up no fewer than 15,000 fossil fish, today housed in the collection of the Miguasha Natural History Museum and in the collections of over thirty leading museums and research centres around the world.

For all of these people, the Miguasha cliffs have been and remain an immense open book on the life and history of our

T he Miguasha Museum's friendly staff pass knowledge on to the public...

planet. Paleontologists of yesteryear and today have explored the fossils in their tiniest details, trying to squeeze out every drop of information, however minor, in an effort to understand not only the fishes and plants but also their environment and the Devonian geological period. The Escuminac Formation still has enough to keep several generations of researchers busy! New species remain to be described. New specimens remain to be discovered so that we can better understand their ecology, morphology, and evolutionary position in the great tree of life. Such has been the case with *Elpistostege* for example.

... but there is also much to learn "in the field," amid the rocks.

Photo : APM

More than ever, the Miguasha cliffs are playing a key role in our understanding of life on this planet. Its fossils have borne witness to several major stages of evolution, such as the origin of the first forests with *Archaeopteris*, the diversification of terrestrial arthropods with the scorpion *Petaloscorpio*, the extinction of two groups of jawless fishes—the anaspids and the osteotracans, the ori-

Norman Parent, a technician at the Miguasha Natural History Museum, preparing a fossil fish specimen in a laboratory with a compressed-air chisel and under a stereoscopic lens. At times, one may need up to hundreds of hours to remove the sediments that cover a fossil.

Photo : APM

Right-hand page: the Natural History Museum on Miguasha Point.

Photo : Armand Dubé

gin of the coelacanths with *Miguashaia*, and the origin of the tetrapods with *Eusthenopteron* and *Elpistostege*.

Few paleontological sites in the world offer as comprehensive a view on the past as the Miguasha site. Its animal and plant specimens, preserved as microfossils or macrofossils, offer a cross-section of the water and land ecosystems. The Miguasha site is not just a collection of fossils but a window in time on life as it was 370 million years ago.

The quality of preservation of the specimens from the Miguasha cliffs has undeniably made interpretation easier for researchers. To carry out this work, paleontologists now resort to a wide range of sciences: taphonomy, palynology, sedimentology, geochemistry, biostatistics, cladistics, and comparative anatomy. These sciences give us specialized insight into the processes of fossilization, the characteristics of microorganisms, the formation of sedimentary rocks, the chemistry of rocks, the statistical study of living things, and the kinship between fossils and living things. Such areas of specialization open up unlimited horizons for men and women of science, both today and tomorrow, the ultimate objective being to understand the past better, to explain the present, and to peek into the future to grasp the continuity of the evolution of life on our planet.

On December 4, 1999, Miguasha Park was added to the list of UNESCO World Heritage Natural Sites, thus crowning our story of people and fossils. This prestigious recognition, however, may be just one step among many. The Miguasha cliffs still have a lot to tell.

As the Canadian biologist Pierre Béland said, *"For anyone who venerates the Earth, Miguasha is a little like one of those sacred sites where primitive men placed the fount of their myths and which they didn't dare destroy, for fear that the great order of things be forever upset. The onus is on us to preserve it for future generations."*

Suggested Readings

Barrette, C., *Le miroir du monde*. Éditions Multimonde, Québec, 2000.

Bourque, P.-A., *Planète Terre* (WebSite), 1999. http://www.ggl.ulaval.ca/personnel/bourque/intro.pt/planete_terre.html

Rondot, J., *Les impacts météoritiques à l'exemple de ceux du Québec*. MNH, Québec, 1996.

Specialized Readings

Le Guyader, H. (Dir.), *L'évolution*. Bibliothèque Pour la Science, 1998.

Long, J. A., *The Rise of Fishes - 500 million years of evolution*. The Johns Hopkins University Press, Baltimore, 1995.

Maisey, J. G., *Discovering Fossil Fishes*. Henry Holt and Company, New York, 1996.

Paxton, J. R. and W. N. Eschmeyer (Eds), *Encyclopedia of Fishes*. Academic Press, San Diego, 1995.

Schultze, H.-P. and R. Cloutier (Eds), *Devonian Fishes and Plants of Miguasha, Quebec* / Verlag Dr. Friedrich Pfeil, München, 1996.

Printed by

J. B. DESCHAMPS INC.

PRINTERS SINCE 1926